Home Inspection Handbook

by
John E. Traister

Craftsman Book Company
6058 Corte del Cedro / P.O. Box 6500 / Carlsbad, CA 92018

National Electrical Code® and *NEC*® are registered trademarks of the National Fire Protection Association, Inc., Quincy, MA 02269

Looking for other construction reference manuals?
Craftsman has the books to fill your needs. **Call toll-free 1-800-829-8123** or write to Craftsman Book Company, P.O. Box 6500, Carlsbad, CA 92018 for a **FREE CATALOG** of over 100 books, including how-to manuals, annual cost books, and estimating software.
Visit our Web site: http://www.craftsman-book.com

Library of Congress Cataloging-in-Publication Data

Traister, John E.
   Home inspection handbook / by John E. Traister
     p. cm.
   Includes index.
   ISBN 1-57218-046-3
   1. Dwellings -- Inspection -- Handbooks, manuals, etc. I. Title.
   TH4817.5.T73   1997
   643'.12--dc21   97-13999
                                        CIP

©1997 Craftsman Book Company
Second printing 2001

Contents

Preface 5

Chapter 1   The Home Inspection Business 7
Getting Organized, The Field Inspection, Meet the Owner, Energy Considerations, The Home's Utilities, Problem Areas

Chapter 2   Building Sites and Landscaping 39
Restrictions on Land Use, Evaluating Homes, Landscaping, Inspecting Landscaping, Inspection Checklist — Building Sites

Chapter 3   Foundations 57
Footings and Walls, Concrete Slabs, Inspecting the Foundation, Inspection Checklist — Foundations

Chapter 4   Building Structures 71
Wood Construction, Masonry Structures, Inspecting Building Structures, Inspection Checklist — Structures

Chapter 5   Roofing 87
Roof Styles and Design, Roofing Material, Rafters, Inspecting the Roof, Inspection Checklist — Roofing

Chapter 6 Chimneys and Flues.................................. 113
*Chimney Inspection, Fireplace Inspection, Inspection Checklist
— Chimneys and Flues*

Chapter 7 Interior Finishes.. 127
Inspecting the Interior, Inspecting Trim and Built-ins, Ceilings

Chapter 8 Exterior Finishes... 151
*Wood Siding, Miscellaneous Wall Siding, Masonry Facings,
Surface Coatings, Inspecting the Home's Exterior, Inspection
Checklist*

Chapter 9 Insects, Vermin and Decay 171
*Termites, Vermin, Wood Decay and Radon Gas, Inspecting the
Home for Pests and Rot, Inspection Checklist*

Chapter 10 Electrical Systems 187
*The Electrical System, The National Electrical Code,
Inspecting the Electrical System, Inspection Checklist*

**Chapter 11 Heating, Ventilating and
Air Conditioning** 227
*Types of Heating Systems, Basic Air-Conditioning Concepts,
Inspecting the HVAC System, Inspection Checklist*

Chapter 12 Plumbing Systems 253
*The Basic System, Hot Water Systems, Indirect Water Heating,
Inspecting the Plumbing, Inspection Checklist — Plumbing*

Chapter 13 Operating Techniques 275
*Checklists for Going Into Business, Getting Started . . . Right!,
Cost Accounting, Public Relations, Advertising*

Chapter 14 Preparing Reports....................................... 297
Types of Reports, Practical Applications

Index 317

Preface

Home inspection encompasses a fascinating world with many opportunities. Few businesses or professions offer as much opportunity, with minimal investment, as home inspection.

A little more than a decade ago, fewer than 5% of the homes sold in the United States were inspected or appraised by professional home inspectors. Most lending institutions required only that an officer of the firm briefly look at the property and note its general condition. However, by 1990, over 35% of home mortgage loans required some form of inspection. Within five years (by the year 2000), it is estimated that 90% of all federally-insured lending institutions will require a complete inspection before granting any loans.

Inspection requirements became mandatory because of the economic issues facing the United States in the 1980s and the financial industry's dilemmas. Problems in the savings and loan industry forced bank examiners to scrutinize every loan record and application. Most banks adopted their own policies requiring approved inspectors and appraisers to carefully examine property for all home mortgages. Insurance companies have recently required an inspection of property before they will issue policies. Even home buyers, wary of rising costs and rates, hired home inspectors to thoroughly investigate property before making a purchase.

Thousands of existing homes are sold each year. Billions of dollars are spent on new construction that takes place on a continuing basis.

Thousands of loans are refinanced each year. Most of these transactions eventually need the services of a professional home inspector.

Many home inspectors also appraise all types of property — rural property, tract developments, and commercial property. Doing so broadens the home inspector's opportunities for profitable employment with banks, real estate firms, savings and loan companies, government-guaranteed loan organizations, and others.

This book is designed to review the way buildings are designed and constructed, which areas of buildings should be inspected, and how to inspect them. Suggestions are then given to help prepare an inspection report; reports designed to please and meet the specifications of lending institutions and other organizations requiring home-inspection services. In fact, the data presented in this book will prove invaluable on practically every home inspection project encountered.

John E. Traister
1995

Chapter 1
The Home Inspection Process

The best way to organize an on site inspection is to use a checklist. Such lists must remain flexible and should be amended each time an inadequacy is discovered in the field. This book has included various checklists for your use as guides. However, they are not gospel. They should be amended and customized to suit your particular circumstances.

Some of the information included in this chapter touches upon detailed information outlined in other chapters. It is designed as an overview of home-inspection procedures required on each job, and to help you visualize a sense of movement through a building and on the surrounding grounds. As a complement to this, partial checklists have been developed to help in organizing your own procedures. Additional chapters will reinforce these inspection techniques.

GETTING ORGANIZED

One of the key words for a successful inspection business is *organization*. While this chapter material is indeed helpful in learning the fundamentals, it is no substitute for the personal intuition and working methods that come through experience. This book will lay a solid foundation for you. It will then be up to you to apply what you have learned ... the checklists contained herein are one of the "tools" that will help you. You should customize your own lists in a way that will be comfortable for you and fit your working method as well as the particular clientele that you are marketing.

The Job Checklist

When a client calls, or you make a contact from a referral, get the correct and complete name, address and phone. This is basic. Pertinent data might include the following:

- Date

- Name

- Address

- Phone

- Billing address

- Address of property to be inspected

- Present owner

- Directions to site

- City/County zoning of site

- Your proposed method of billing (hourly or by job)

You should also be prepared to give potential customers the approximate amount of your charges before beginning the inspection. The fee that you normally charge may differ greatly from what the customer thinks the job is worth. An inspection-form head should look approximately like the one in Figure 1-1.

```
  Name_____

  Address_____

  City_____State____Zip_____

  Phone_____

                    **Property Location**

  _____

  _____

  This is my report of a visual inspection of the readily accessible areas
  of this building. Please read the REMARKS printed on each page and
  call for an explanation of any aspect of the report, written or printed,
  that you do not understand.
```

Figure 1-1: A sample home inspection report head.

Site Appointment

The following information should also be obtained at an early date:

- Date of appointment

- Time of appointment

- Who will attend

- Address

- Approximate mileage from inspector's office

- Equipment to be taken to site

All of your checklists are particularly important when discussing a job with a client, either over the phone or in your office. Referral to a checklist will show the client how thorough and organized you are — giving him or her confidence in hiring you — much the same as an airline pilot goes through an extensive checklist before each and every flight.

Inspector's Equipment

An inspector will need to have various pieces of equipment in order to properly evaluate a building. A level is a necessity. Look to purchase a high-quality level that is short enough to carry conveniently. Keep in mind, however, that the shorter the level, the less accurate it is. Levels can be quite sophisticated; some even have a digital readout of degrees slope. More than likely, a simple bubble level will suffice. See Figure 1-2.

Various electrical testing instruments will be essential. Your basic testing instrument should be a voltage tester, used to determine what wires or equipment are "live." Many home inspectors purchase a combination volt-ohm-ammeter such as the one shown in Figure 1-3.

Measuring tape of good quality with a capacity of at least 100' is recommended (don't use electronic sonic devices — keep things sim-

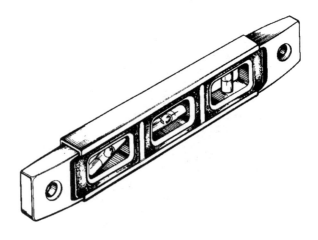

Figure 1-2: A small bubble level will usually suffice for most home inspection jobs.

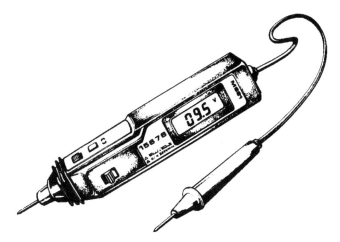

Figure 1-3: An electrical testing instrument is essential for testing electrical circuits and apparatus.

ple and use a manual tape measurer). A shorter tape measure (Figure 1-4) that may be carried on your belt is also highly recommended for measuring shorter distances such as the width of doors and windows, depth of cabinets, ceiling heights, and the like.

A flashlight is absolutely necessary for all home inspection jobs. In fact, you should have at least two of them, along with spare batteries

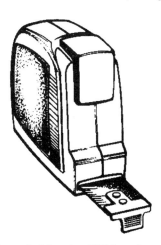

Figure 1-4: You will need at least a 100' tape to measure the outside perimeter of the home, but a shorter tape that can be clipped to your belt will also be handy for the majority of your measurements.

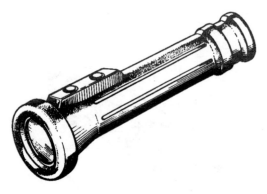

Figure 1-5: Inspectors should carry two flashlights, as well as extra batteries and bulbs for each.

and bulbs. A flashlight is necessary for inspecting under crawl spaces, under porches for structural defects or insects, in dark corners of basements at an electric panel, and a host of other uses. You don't need an extremely powerful one; the one shown in Figure 1-5 should be adequate.

Always keep a complete set of screwdrivers in your car. In general, a small and medium screwdriver with wedge-shaped blades and one Phillips head screwdriver will suffice for most of your needs. The ones shown in Figure 1-6 are good choices.

The medium-size screwdriver will be used for removing panel covers on appliances and electric panels. It is also good for poking into suspicious wood members for signs of dry rot and termites — although

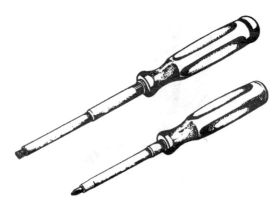

Figure 1-6: Screwdrivers will be useful on every job. You might also want to carry an ice pick for probing into suspicious timbers for decay or termite damage.

many home inspectors prefer an ice pick or scratch awl for this latter operation. The smaller screwdriver is useful for removing the plates from wall switches and duplex receptacles so that the wiring and devices may be checked. You will also find many other uses for a set of screwdrivers.

Other necessary tools include an inspection mirror, a dial thermometer, binoculars (for a close-up look at certain parts of the house, like the roof, that are difficult to get near), and a ladder (a folding 6' step ladder will serve most of your needs). Of course, you should also have at least two pens (black or blue), several business cards, paper for taking notes, graph paper for making sketches of the home's floor plan, and a clipboard.

If you have agreed to perform a radon test, make sure that you have some radon detection units in your car. Spot testing can be done with either a charcoal canister or an alpha track device (the lab test costs between $25 and $50.) Get more information from the National Association of Home Builders (NAHB) and the Environmental Protection Agency (EPA) of the Federal government. Usually the charge for the radon test is added to the cost of the inspection.

There are other tools that will come in handy from time to time. For example, an inexpensive compass will indicate in which direction the house is facing; gloves to protect your hands from dirt and splinters; roofers boots (non-slip soles) if you are required to walk on the building roof; camera and film to record the appearance of the home, including close-up details. Here's our job-site equipment checklist so far (again, revise it to suit your needs):

- Radon detector

- Screwdrivers

- Electrical testing instruments

- Moisture meter

- Level

- Tape

- Calipers

- Flashlight/miners hat light

- Magnifying glass

- Binoculars

- Profile gauge

- Gloves

- Camera and film

- Roofers boots (non-slip soles)

- Fold-up ladder

- Compass

- Hand mirror

A brief description of some these tools is in order. A moisture meter is useful for determining the moisture content of timber and in areas of the house that seem to be overly damp. A magnifying glass can be used (for one) to distinguish between termites and flying ants. The hand mirror, used in conjunction with a flashlight, can be used to explore the inside of wall partitions.

Dress

Home inspectors are in constant contact with the public. To be successful, always present yourself as a professional. Nothing projects this image better than your appearance. When you show up for the inspection, be sure that you are neat and clean. If you get out of your car in jeans and a T-shirt, your client will think that you don't take your job seriously. Appropriate attire for men include slacks, a sport coat, a dress shirt, and a tie. Women should wear slacks, a blouse, and a blazer. To allow more freedom for movement, you can remove your coat when you begin the inspection.

Wear flat shoes with a rubber sole (not tennis shoes). The rubber sole is necessary for safety. It will provide traction and help prevent electrical shock. Keep your shoes clean and polished. Keep a raincoat and a set of coveralls stored in your car. Store them in a place where you won't stack anything on top. A wrinkled or stained raincoat or coveralls do not present a professional appearance.

THE FIELD INSPECTION

As you approach the home that you will be inspecting, observe the lay of the land. Is the area on high ground or does it seem to be low-lying? This may seem like an insignificant question, but sites on high ground will generally have fewer storm-drainage problems and therefore will tend to have usable dry basements. The ideal location for a house is on a large knoll on the lot with drainage leading away from the house in all directions. See Chapter 2. Note the spacing of street storm drains (usually built adjacent to the sidewalk curbing in the street). If there is a lot of dirt, leaves and dirty sticks and trash at the drain, there might be a problem with poor drainage or a drain that backs up into the street. This may be the first indication of problems at the building site.

If inspecting a rural home, look for low-lying swampy areas where water might be standing, or where water stands most of the time. These areas can be detected by thin or non-existent grass growth, mud holes, or a brownish-gray discoloration of the surrounding low-lying plants. Willow trees at the site also indicate high ground moisture.

House Style and Roof

Stand back and look over the house. Take photos of the main house, carport, garage, and all outbuildings. You will want to examine each of these if requested. Large trees overhanging the house should be noted as possible lightning/wind damage threats to the home. At this point you should indicate the house style (Figure 1-7 on the next page), type of roof, placement of house on building lot, type of roof covering, skylights (Figure 1-8); flashing for the roof (Figure 1-9), and other

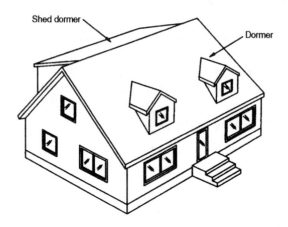

Figure 1-7: Identifying a house or roof style is not always straightforward. The house shown may be called a modified Cape Cod with two dormer windows in front and a shed dormer in the rear.

general conditions as indicated in the partial form in Figure 1-10 on page 18.

Observe the general condition of the exterior facings and the condition of the caulking and paint. Look for areas that might possibly admit rain water. Look for signs of rot or decay or areas where the paint has failed. Observe water stains on the wall and roof surface. Directly below these might be trouble spots for cracked foundation walls or basement leaks.

Make a note as to the orientation of the house. In some locations the

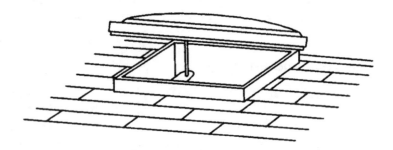

Figure 1-8: Note all roof openings, including skylights, vents, and all roof protrusions such as plumbing VTRs.

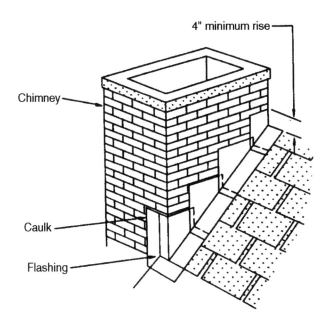

Figure 1-9: While inspecting the roof, check the flashing and caulked joints. Faulty flashing will leak and damage structural framing and interior finishes.

North side may weather much more rapidly than the rest of the house. The West and North sides of the roof depending on your location may weather out faster too. When you get to the inside you may find that visible traces of roof leaks such as water stains are in the North and West rooms of the house. These observations may not be of any immediate consequence but are things that the client can begin to anticipate about his new building and give him confidence in the inspector's observations and thoroughness.

If you choose to start the inspection on the outside, put your ladder in position to get you to the roof or wall areas needing close examination. Give special attention to roof valleys where one surface is flashed onto another. Check all flashing, including that around the chimney. Note whether the chimney flashing is "stepped" or runs parallel to the slope of the roof. If it is not stepped and flashed back into the brick joint then the flashing is placed next to the face of the brick chimney and caulked at the top. After a time this caulk will fail. Leaks generally begin at flashings and at caulked joints. Beginning on the exterior may give you some important leads to problem areas on the inside. Simple

House style_____						
Is the house situated on an elevated part of the lot for good drainage?					Yes	No
Type of roof:						
Gable	Gambrel	Hip	Flat		Shed	Other
Type of roof covering:						
Asphalt shingle	Tile	Metal	Fiberglass		Roll roofing	Other
Type of roof flashing	Tin	Copper	Other_____			

		Yes	No
1.	Are there any dormer windows?	_____	_____
2.	Are there any loose shingles?	_____	_____
3.	Are there any smooth spots where the roofing material has worn away?	_____	_____
4.	Is every opening in the roof (skylight, chimneys, vents, etc.) properly flashed?	_____	_____
5.	Are any boards or shingles on dormer windows loose or in need of repair?	_____	_____
6.	Does the roof sag?	_____	_____
7.	Are the soffits, eves, etc. in good condition?	_____	_____

Overall condition of the roof and related structure:	Good	Fair	Poor
Items in need of repair:	1.		
	2.		
	3.		
	4.		
	5.		
	6.		
	7.		
	8.		

Figure 1-10: The inspection checklist continues with building site, house and roof style, and general conditions of the roof.

The Home Inspection Process

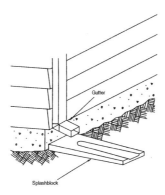

Figure 1-11: While inspecting the roof, check the gutters and downspouts. See if the downspouts have splashblocks to divert the water away from the foundation.

things which are often overlooked are the checklist items which you want to zero in on. For instance, observe the ground immediately adjacent to the house. It should slope down away from the building. If it is level, try to determine where rainwater flows.

Take notes on the downspouts. See Figure 1-11. With experience, one can usually tell if the gutters and downspouts are too small. Maybe there are not enough downspouts in which case the gutter water may overflow at one end; look for water stains. The foundation at one point may be subjected to an inordinate amount of water leading to soft ground bearing conditions and failure of the foundation. If gutters are bent or warped out of alignment, ice may be accumulating in them causing them to sag. This condition is caused by blockage of the downspouts. If they are poorly fitted together, leaves and ice may accumulate causing blockage.

Foundation

While on the outside of the house, and after checking the roof area, you should examine the foundation closely. Start where the foundation walls are exposed. Cracks (Figure 1-12) may not indicate dangerous structural problems but they will usually leak and cause differential settlement of the foundation. Trees located too close to the building will cause cracks. Root systems can exert tremendous pressures over

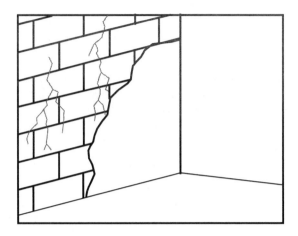

Figure 1-12: Cracks in basement walls do not always indicate serious structural defects, but they will almost always leak.

time. A large dead stump adjacent to a foundation wall crack will show that even though damage has been done the culprit has been stopped.

A basement or crawl space is always the lowest level of a house and parts of it is usually left unfinished by the builder. Even if the owners have a finished recreation room, and perhaps a finished laundry room, there will almost always be some sections left unfinished; that is, the furnace or utility room, storage space, and perhaps a basement garage. If you are inspecting a house that has unfinished areas, the exposed foundation walls, overhead floor joists, and other exposed parts of the house can tell you a lot about the structural soundness of the house. This not only applies to the foundation, but also to the wood framing as well as the electrical, plumbing, heating ventilating and air conditioning systems. Parts of each will usually be exposed in unfinished basement areas.

Crawl spaces offer the same advantages; that is, parts of the floor joists, electrical wiring, plumbing pipes, etc., normally can be seen from the crawl space of any house. Building codes require that the crawl space have at least one entry for repairs, maintenance, or for fighting fires. This is your access point to examine the crawl space of homes that you are inspecting. Therefore, make certain that you have a good flashlight when inspecting such areas. Although the National Electrical Code now requires at least one lighting fixture — with an ON/OFF switch — in every residential crawl space, there is always the chance the lamp will be defective. Furthermore, always carry a pair of coveralls in your car, and also some slip-on rubbers to protect your shoes. A crawl space is just what the name implies; that is, you're going to have to crawl on your hands and knees to move around in the area.

Examine behind bushes and plants adjacent to the house. Examine all porches, soffits, and crawlspaces. Be prepared to crawl into shallow crawlspaces. These areas may never have been seen by the owner and may be surprised as to your findings. Be very careful of electric lines. Observe the presence of rodents and any animal or insect activity. Run through a complete crawlspace checklist which could be similar to the basement checklist.

Since the foundation is one of the most important areas of the home, the home inspector should spend considerable time looking at the foundation and its related components. Remember, the remaining building structure is only as good as its foundation. Check the humidity level and look for signs of water infiltration. It might be helpful to check on any ground water problems in the area. Also note the basement's headroom and its future limitations. Many homeowners like to utilize the basement area for recreation, workshop, or laundry area. Is the basement suitable for these?

A checklist for basement and foundation inspections is shown in Figure 1-13 on the next page.

Structural Members

A home's structure goes hand in hand, in importance, with the building's foundation. Defects in either can result in serious problems that will greatly affect the overall value of the home.

A broken window pane or door is minor. Both can be repaired or replaced with relatively little cost. However, when it comes to the building's structural framing, we enter into an altogether different situation. Framing corrections can run into the tens of thousands of dollars, and can make a tremendous difference in the value of a home.

Since most framing is hidden by sheathing or other types of finishes, most structural framing in existing homes is not readily visible to the home inspector. However, there are still many ways in which this part of the home may be inspected. Once you know what to look for, and gain experience in these areas, you should have little difficulty in detecting any structural deficiencies and filing your reports accord-

Type of foundation:

Block___	Reinforced concrete___	Slab on grade___	Other___
Full basement___	Partial basement___	Crawl space___	

FOUNDATION AND BASEMENT CHECKLIST

		Yes	No
1.	Are the foundation walls free of vertical cracks?	___	___
2.	Are the piers in the crawl space or basement free of cracks?	___	___
3.	If cracks exist, are they hairline cracks or V-cracks?	___	___
4.	Does the crawl space have adequate ventilation?	___	___
5.	Are the walls straight; that is, no bows or obvious curves:	___	___
6.	Does the house have drain tile around the perimeter of the foundation?	___	___
7.	Are vapor barriers properly installed?	___	___
8.	Does the house smell clean (not musty)?	___	___
9.	Are the basement walls dry?	___	___
10.	Does the slab floor feel dry?	___	___
11.	Are there any large trees close to the house whose roots may damage the foundation or footing?	___	___
12.	Are there adequate floor drains in the basement floor?	___	___
13.	If a sump pump is used, does the basement floor have the correct taper for adequate drainage?	___	___
14.	Are there any signs of settlement (sunken floor, cracks in walls, floor not level)?	___	___
15.	Are there signs of infestation?	___	___
16.	Does the basement have an outside entry?	___	___

Figure 1-13: Inspection checklist for residential foundation and basement areas.

ingly. Often, your only access to the structure is in the attic or an unfinished section of the basement.

Check to see if the walls and ceilings are straight, not bowed. In unfinished basements and attics, check the exposed structural members for cracks and splits.

A structural inspection checklist is shown in Figure 1-14 on the next page. Again, use this as a guide only. Make modifications as necessary and as your work dictates.

Outside Coverings

All painted surfaces begin to deteriorate immediately after installation due to exposure to ultra violet radiation and the elements. Sun screens, anti-oxidants and inert pigments in the paint slow the process, but eventually the vehicle breaks down and the surface becomes dull.

When inspecting outside wall coverings, look for surface deformation, staining, cracking, separation of layers, and separation of the coating from the base. If any of these defects are found, you should also try to determine the reason.

Deformation: This condition is usually caused by the paint being improperly applied; that is, in direct sunlight that causes the solvents in the covering to evaporate too quickly. The same condition occurs, however, if the coat of paint was applied too heavily. The result is wrinkles in the finish coat, changing the reflective value of the paint and also tending to get dirty quicker.

Stains: The most common stains found in exterior wood siding will be from rusted nails, but there are numerous other stains: sap or resin deposits, mildew, soil stains, and a host of others. None of these usually seriously affect the home's overall condition; merely the appearance. Such stains should, however, be noted on your report.

Cracks: Hard, brittle paint is usually the culprit. Old, hard paint is no longer able to respond to the dimensional changes in layers of old finish or the siding underneath induced by changes in temperature and humidity.

Peeling: Paint peeling is due to poor adhesion, causing the layers to separate. A new coat of paint may hide the defect temporally, but if the new coat does not have a way to bite into the old finish, it will soon peel also. The only proper way to correct such a condition is to scrap or otherwise remove the poor layers underneath, properly prime the

STRUCTURAL CHECKLIST			
		Yes	No
1.	Are the wood beams and surfaces free of termites and wood decay?		
2.	Are the ceilings level?		
3.	Are the walls straight; that is, no bows or obvious curves?		
4.	Are the floor joists adequately bridged?		
5.	Have any partitions been removed?		
6.	If the answer to 5 is yes, were provisions made for adequate support of the framing members above?		
7.	Are there any cracked or split joists, trusses, or rafters?		
8.			
9.			
10.			

Overall condition of framing:	Good	Fair	Poor

Items in need of repair:	1.
	2.
	3.
	4.
	5.
	6.
	7.
	8.
	9.

Figure 1-14: Structural checklist.

wood or metal, then apply a base coat. Finally, apply the finish coat at the correct temperature and out of direct sunlight.

Many experienced home inspectors can tell a lot about the entire home by the condition of the home's exterior. For this reason, the exterior is usually the place to begin with any home inspection, and it should be thorough.

The checklist in Figure 1-15 on the next page will assist you in checking many exterior elements of the home.

The Home's Interior

The inspection of the home's interior includes such items as walls, ceilings, floors, steps, stairways, balconies, and railings. All cabinets and counters in the kitchen or bath also fall under this phase of the inspection. Windows and doors are another area that should not be overlooked, as well as the lighting fixtures around the home.

As you enter the home, operate the entry door and examine the hardware (hinges and locks). Check the wall lighting switches and also note if the foyer lighting fixture is in good condition.

Continue through the home observing the walls, ceilings, and floors of each room. Also check and operate all doors and windows, including closet doors. As you enter each room, operate the light switch to make sure it functions properly. Also note the condition of any permanently-mounted lighting fixtures. Table lamps and other portable lights are not the home inspector's concern.

Identify the floor, ceiling, and wall finishes. That is, is the floor carpeted, hardwood, or some other finish? How about the walls? Are they plastered, drywalled, or paneled? If some walls have wallpaper covering, this should be noted in your report.

Look for cracks in the ceiling and walls; also stains, water damage, powdery areas, broken out plaster, walls not plumb or sags in ceiling. Popped nails in the plastered or drywall wall or ceiling could indicate a structural defect.

When checking the floors, look for areas that are not level, damaged flooring, bumps in the floor, warped or loose floor boards. Also identify any floor covering, such as linoleum, tile, slate, and the like. You can carry a tennis ball with you and roll this across floors to see if they are level. A conventional carpenter's level may also be used.

EXTERIOR FINISHES

Type of siding:

Brick veneer_____ Clapboard_____ Wood panels_____ Other_____

		Yes	No
1.	Is the brick in good shape?		
2.	Do all mortar joints fit tightly with no leaks?		
3.	Are the walls straight?		
4.	Are all sections of the siding firm?		
5.	Is the siding in good condition?		
6.	Are there any cracks or chips in the walls?		
7.	Has the siding been nailed properly?		
8.	Are the gutters and downspouts adequately sized and in good shape?		
9.	Are any nails showing through the siding or outside walls?		
10.	Are the exterior doors in good condition?		
11.	Does any door squeak when opened?		
12.	Is the door hardware (hinges and locks) in good condition?		
13.	Do all windows operate satisfactorily?		
14.	Are all outside lighting fixtures in good condition?		

Overall condition of exterior finishes: Good_____ Fair_____ Poor_____

Items in need or repair:	1.
	2.
	3.
	4.
	5.
	6.
	7.
	8.
	9.

Figure 1-15: Exterior inspection checklist.

You should identify the door types; that is, panel, flush, louver, sliding, and the like. Also identify the material from which the doors are made. Then check their condition for:

- Operation.

- Sticking doors.

- Noisy doors.

- Cracks or broken out corners.

- Hardware on each door.

- Scratches.

- Are the door frames square?

Treads on stairs should be solidly fastened; risers should be the same and unbroken. They must also be well lighted so there is no danger of tripping when they are in use. Don't forget understair closets during your inspection.

Identify window types, including the type of glass used. If insulating glass is used, make a note of this. Storm windows on the outside of the home usually indicate that conventional glass is used rather than the insulating type. Don't forget to check the caulking on the outside of each widow for peeling and cracking. Improper caulking will allow moisture into the home, not to mention the heat loss and heat gain through air infiltration.

The checklist in Figure 1-16 will help you not to overlook any important item during your inspection of the home's interior.

MEET THE OWNER

Contact the owner to see what he can or will tell you first hand. The owner may attend your inspection. Try to obtain as much information as possible from him or her. Assure the owner that you will be just as glad to point out advantages of the house as well as any defects. Make every attempt to gain the owner's confidence. Be truthful with the owner. Some information with which he can help:

- How long has the house been occupied?

- What major repairs has been made?

- What appliances are new?

- What are existing problems?

- Ask about basement dampness and drainage of property.

- Does the plumbing system work well?

- Where does the sewer line to the street run?

- Where does the domestic water enter from the street?

- If in county, locate well, cistern, and septic system.

- Find the gas and/or water meters.

- What is the heating bill?

- Has there been termite treatment? (The inspector should read up extensively on this subject. Start by visiting the local library or extension agent. Learn how to spot the signs of termites).

- Has any improvements been made that may not technically meet code requirements?

ENERGY CONSIDERATIONS

All new housing today must meet building code standards for energy efficiency. Older homes had no energy criteria because the cost of heating a home was minimal and there was no cooling. But in 1973 came the first Arab oil embargo and energy costs soared.

The Home Inspection Process

INTERIOR FINISHES			
Type of floors:	Hardwood	Carpet	Tile
	Other:		
Type of walls:	Plaster	Drywall	Panel
	Other:		
Type of interior doors:	Panel	flush	Standard
Number of outside entrances:			

		Yes	No
1.	Is the floor finish in good condition?	_____	_____
2.	Are the floors level?	_____	_____
3.	Do the floors squeak?	_____	_____
4.	Are there any loose floorboards?	_____	_____
5.	Is the floor warped?	_____	_____
6.	Are there any cracks or chips in the walls?	_____	_____
7.	Are there any water stains on the walls or ceilings?	_____	_____
8.	Are drywall joints in good condition?	_____	_____
9.	Are any nails showing through the ceiling or wall finishes?	_____	_____
10.	Are the interior doors in good condition?	_____	_____
11.	Does any door squeak when operated?	_____	_____
12.	Is the door hardware (hinges and locks) in good condition?	_____	_____
13.	Do all windows operate satisfactorily?	_____	_____
14.	Are all lighting fixtures in good condition?	_____	_____

Overall condition of interior: Good_____ Fair_____ Poor_____

Items in need or repair:	1.
	2.
	3.
	4.
	5.

Figure 1-16: Interior inspection checklist.

The inspector must be knowledgeable enough to be able to analyze the energy deficiencies of a home and be able to suggest practical methods by which an older home can be retrofitted for energy efficiency. The word "practical" is important because at some point an energy deficient home may simply have to remain energy deficient because of the extreme labor required to update.

The inspector must become familiar with the Council of American Building Officials' (CABO) *Model Energy Code*. This code, which is updated yearly, has become the guide for most new home construction in the country. Become familiar with the R-Value zone in your area. This is the temperature zone (roughly running latitudinally from the East Coast to the West Coast) requiring insulation that has the ability of slowing the movement of heat to a certain established level known as the "R-value". Refer to the code under your zone to find the appropriate ceiling, wall and floor R-values.

It may be difficult to determine how well a house is insulated. Check every possible access to the stud space. You may have to get creative. At some locations your only clue to wall insulation may be from behind the electric outlet cover. In the attic and in the basement insulation may not be covered. A full inspection may be possible here. Make note of the neatness and care in which the insulation was installed in addition to the type and thickness. If a client wants a full report on all of the insulation in the house you must be direct with him and state that you are able to report on only those areas or spots which are accessible. Be careful about drawing too many generalized conclusions based on spot checks or on only those areas that are visible. This applies to most other components of the house as well.

Storm windows will indicate a plus for the energy side. See that they are tight fitting and can readily be raised and lowered. Note the condition of the paint of the house windows. If the paint has failed the storm windows were installed without properly maintaining the house windows. The new owner may have to remove all of the storm windows to paint and repair.

Attic

A client may come to you for advice on structural matters. He may have decided even before buying a property that it will be perfect for his family if he can remodel the attic for bedrooms. Of course, you must refer him to an engineer (it will be good to establish ties with a structural engineer for other questions that you may encounter in your business). You can, however, develop a feel for the potential problems that may throw a damper on the client's visions. Let him know that an attic floor may not have been designed to meet building code loading requirements for occupied rooms such as bedrooms. Also advise him of potential headroom problems due to rafter "collar" ties or other headroom infringements that may need analysis by the engineer to determine the feasibility for the space's future use. Note also that in order to put a cathedral type ceiling in the attic there is the condensation factor to consider. If placed between rafters, insulation should be spaced at least an inch down below the bottom surface of the roof sheathing.

THE HOME'S UTILITIES

The home's utilities include the electrical, plumbing, heating, ventilating and air conditioning systems. All of these systems should be carefully inspected because all of them directly influence the overall condition of any structure.

Plumbing

It was common years ago to vent the house plumbing system directly into the chimney within the attic area. While it is not currently permitted by the National Plumbing Code, you will find this situation in older homes. If found, check the condition of the vent, its seal to the chimney and its noncombustibility.

The only area in which plumbing piping is partially visible is in the basement. Therefore it is next to impossible to visually inspect the complete plumbing system. If the house is new then the local building inspector will have done this already. That is his job. If something major about the house seems out of line to you then it may be possible to speak to the city/county inspector. This would be a very unusual situation. It would be advisable for a professional inspector to become familiar with all of the duties of the local building inspector. This will enable him to throw off all of the unseen and impossible-to-detect items to the shoulders of the city inspector. He was required to inspect those items as the house was built. The professional inspector must not set himself up as being able to verify the correctness of those items. It might be useful to note in your report to your client those for items which the code inspector is accountable.

The inspector should note any plumbing items which seem out of line and which (even though they can't be inspected because they are hidden) may potent greater consequences. For instance, the inspector should know basic plumbing pipes and their acceptability. Hot and cold water supply can be distributed by copper pipe, galvanized steel, CPVC, polybutylene. Plastic pipes are usually marked as to material, temperature and pressure.

Water heaters can be a source of problems for a new landlord. The inspector should note the vintage of the model. Life expectancy of a water heater varies but one that is found to be 15-20 years or older will most likely die sometimes in the near future. Visual inspection of the entire heater may not be practical because of its tight location against the wall, etc., but be sure to check for rusting at the top of the tank since a bad heater usually rusts through at the top. Also note the capacity of the tank. A standard 4500W electric heater should be 30 gallon capacity for 1 or 2 people; 40 or 52 gallon for 3 to 5 people (30-40 gallon for gas heaters). Any large water heater, say, 30 gallons or more, should have 4500W heating elements. Those with a lesser wattage, such as 1500W or 2500W are probably more than 20 years old.

During your walk through the house, turn on the faucets for all sinks and showers. You are checking for *functional water flow*, not for water pressure. Check that the flow is adequate. With the water running, flush the commode. The flow will decrease, but it should not stop. Turn off the faucet and shower before leaving the room. Check the kitchen and utility sinks as well. Also check the caulk around the sinks and tub.

HVAC Systems

The heating plant, shown in Figure 1-17, is usually in the basement, or in the case of a single-floor dwelling, in a utility or furnace room. Note the vintage of the furnace and see that is is properly vented. If the basement has not been an occupied area, determine if the area has enough ventilation. If the furnace or plumbing pipes are corroded badly there may be some ventilation shortcomings that need immediate correction.

Examine the insulation around the furnace, ducts and water lines. This was one of the most common uses of asbestos. Asbestos is light gray in color and was used for insulation around steam and hot-water pipes. If you suspect asbestos insulation, note it on the report. Word the report as to avoid a definite statement: "Material of this type has been know to contain asbestos." Modern day ductwork may be either insulated metal, rigid board insulation formed into rectangular ductwork or flexible insulated round ductwork. Note that all ductwork in an attic must be insulated. The same goes for ductwork in a ventilated crawl space or other nonconditioned space. Under certain circumstances, refrigerant pipe must also be insulated.

The inspector should become familiar with code requirements for flue pipe and chimney connectors. There are lots of technical requirements on flue and chimney venting — too lengthy to consider here. One of the more common venting problems with which the inspector comes into contact is installation of wood burning stoves. Such stoves are very popular these days. Make sure that a stove is not vented into a working fireplace flue. Generally speaking, two or more heating devices should not be connected to the same flue because of the

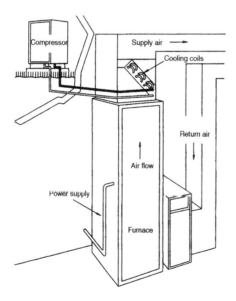

Figure 1-17: Always check the age of the heating and air-conditioning system. A dated HVAC system will certainly need replacing in the near future.

resulting problems with adequate draft. Become familiar with the various types of metal flues used with these stoves and their proper installation. Inspect all pipe or flue connections to the chimney. They should be airtight for proper draft.

Inspect the chimney flue from the roof top. You will need a flashlight for this. If there is a chimney cleanout door in the basement, inspection can be made with the use of a mirror. Also check the heat exchanger or gas- or oil-fired furnaces for cracks and rust. Use a hand mirror and flashlight for this purpose.

An inspector has many responsibilities and is faced with a multitude of technical problems. Though "a little knowledge is a dangerous thing," the inspector will never know it all. Be aware of your own limitations and not blunder into areas beyond your expertise. Knowing when and how to refer the client to other specialists will become critical to a good reputation.

Electrical System

Carefully inspect the electrical system of the house, starting at the service entrance. Check the fuses and breaker box for rust. Be sure that all panelboards and safety switches are firmly fastened to the wall and are located within the code requirements.

Open electrical panelboards, safety switches, and load centers to check the condition of the fuses and/or circuit breakers. Be sure that all "knock-outs" are occupied with a cable or conduit connector or else filled in with a knock-out closure. Carefully remove the cover from the panelboad and check the wiring to the bus bars. *Use extreme caution when inspecting the inside of any electrical box.* Check the connections for tightness. All cables coming into the box should be secured with a cable clamp. Be sure to reinstall the cover.

During your walk through the house, check the operation of all electrical appliances — dishwasher, electric range, washer, dryer, attic fans, etc. Also check every permanently mounted lighting fixture (table lights are not considered part of the house). Use a three-wire circuit tester to check all wall receptacles. Test all ground-fault circuit interrupter receptacles to be sure they disconnect power when tripped. The wiring method used can normally be seen in an unfinished attic or basement area.

PROBLEM AREAS

Just as some homes are inadequately insulated, others are poorly ventilated. Ventilation is especially important in the bathroom, attic, and kitchen. Bathrooms should have mechanical ventilation. If a bathroom is not properly vented, the inspector will be able to spot problems immediately. Wallpaper peeling, paint cracking and deteriorating, along with corroding pipes, metal fixtures, and electric baseboard heating units are usually obvious. Immediately surrounding the shower, the heads of drywall nails will rust through the joint compound. Mildew and mold will accumulate in the grout joints of ceramic tile. Wall tile glued directly to plaster backing will deteriorate in spots and cause loose tile. Moisture will penetrate to the wood joists and studs, causing rot.

The bathroom is a very important room to the homeowner and should be thoroughly examined. If tile work has just been installed, get the owner to tell the client the exact nature of the backing material. Some materials with the tradename of "Wonderboard" and "Durock" are expected to have a long life with high moisture-resisting capabilities.

There should be an adequate air change in all bathrooms. This air change is accomplished either by an exhaust fan or open window. Check to see that exhaust fans vent either to the outside or a well-ventilated attic. Accumulation of moisture in an unvented attic space will rust the structural metal tie plates used on prefabricated wood roof trusses. At the kitchen, an exhaust fan is usually located directly above the range. An approved non-vented hood is acceptable and the fact that it is not vented directly to the outside should be noted in the inspection report. Clothes dryers must be vented to the outside because of the extremely high heat of the exhaust air.

Bathroom floors can also be adversely affected by moisture. Sometimes ¼" plywood is the underlayment for linoleum tile. In order for the ¼" plywood not to buckle with moisture, it must be peppered with nails to the floor sheathing on which it lays.

Careful examination of the plumbing system cannot be overstressed. Turn the sink spigot on at the same time as the tub spigot and then flush the water closet several times. You are looking for an abnormal backup of water or water that may backup in an adjacent bathroom. Older homes were often constructed with exterior sewer (waste) lines of a material called "Orangeburg" pipe. After 20 years or so, this pipe will totally disintegrate if driven over or if roots grow around it. You may not be able to spot this condition. Make sure your report states that you were unable to tell if this type of pipe was installed and that if it was, the owner can expect sewer backups and probably total exterior line replacement to the city connection.

If the water closet doesn't flush completely or it seems to flush sluggishly, then there may be a calcium buildup at the rim supply ports. Jamming the end of a coat hanger into some of the ports may answer this question.

The bathroom is worth its own checklist, which may begin to take the form shown in Figure 1-18. Modify it to include the kitchen if you wish.

The Home Inspection Process

BATHROOM 1

Location:		
Built-in tub_____	Leg tub_____	Stall shower_____
Ceramic tile: In mortar__	In mastic_____	Fiberglass surround____
Plumbing leaks:	Some signs_____	None noted_____
Ventilation:	Fan_____	Window_____
Floor covering:		
Satisfactory:_____	N/A_____	

BATHROOM 2

Location:		
Built-in tub_____	Leg tub_____	Stall shower_____
Ceramic tile: In mortar__	In mastic_____	Fiberglass surround____
Plumbing leaks:	Some signs_____	None noted_____
Ventilation:	Fan_____	Window_____
Floor covering:		
Satisfactory:_____	N/A_____	

BATHROOM 3

Location:		
Built-in tub_____	Leg tub_____	Stall shower_____
Ceramic tile: In mortar__	In mastic_____	Fiberglass surround____
Plumbing leaks:	Some signs_____	None noted_____
Ventilation:	Fan_____	Window_____
Floor covering:		
Satisfactory:_____	N/A_____	

Figure 1-18: The bathroom checklist.

Chapter 2
Building Sites and Landscaping

Many single-family residences are built as part of a subdivision; that is, on a piece of land that has been sectioned or subdivided into two or more building lots.

As seen in Figure 2-1 on the next page, acreage may be subdivided into dozens, hundreds or even thousands of individual building lots, which may be built on immediately by the subdivider/builder or only when a purchaser decides to build.

Many subdivisions are developed on speculation; that is, the owner/builder will build a dozen or so homes on the sectioned acreage and offer them for sale. The buyer is then responsible for obtaining the financing. Before any such loan can be approved, however, the lending institution will often seek the services of a home inspector to insure their investment against faulty construction, and other items that may make the loan a bad investment. When a home inspector is called upon to make an inspection, there are many considerations that the home inspector must take into account: taxes, assessments, easements and other restrictions that might affect the value of property. The building lot or site should be one of the first areas evaluated.

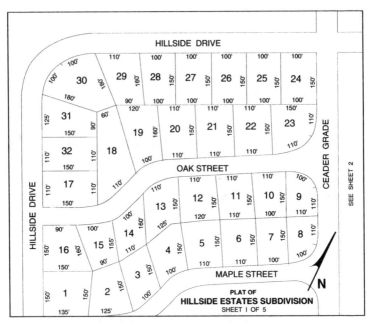

Figure 2-1: Tracts of land may be subdivided into many lots. Some developers build on all lots at once, while others may build a few sample homes, then wait for buyers before further construction is carried out.

RESTRICTIONS ON LAND USE

Not until the twentieth century did cities try to legislate land use. Although multistory tenements and row houses (which we now refer to as townhouses) had city lots crowded for almost a century, population density became critical only with the wave of immigration that began in the late nineteenth century. Their settlement pattern is still evident today. The propertyless and generally penniless newcomers gravitated to the poorest section of a city, crowding many more families into the area than had been there before. As these groups prospered and were able to afford better and more spacious accommodations elsewhere in the city, or even in the suburbs, other groups, equally poor, took their places in what became a new ghetto. Of course, buildings undergo steady deterioration, and the general condition of the neighborhood worsens with each succeeding group.

Since it would be impossible to enforce legislation limiting the number of occupants in a building, the first zoning legislation — in

Building Sites and Landscaping

New York City — sought to limit building density in terms of the size of each building and its placement on the lot, with minimum requirements for front, rear and side yards. By 1930, cities were also regulating land use. Although some separation of residential, commercial and industrial properties had already taken place naturally, as in the case of industries that flourished in areas offering good, cheap transportation and proximity to natural resources, zoning finally enabled specific areas to be designated for only one use, even when other uses would also be suitable.

Zoning Laws

Regulations for controlling the use of land are called *zoning laws*. The purpose of such laws is to protect a homeowner's investment in a new home or land. If a home is in an area zoned "residential," no one can build a factory or other office building or store in the area.

Most communities have zoning laws; the most common zones are residential, multifamily, commercial, light industrial, heavy industrial, and farms. The table in Figure 2-2 shows these zones according to building restrictions. Some municipal governments are also adding a "cluster" or "multiple unit" zone to permit planned unit developments.

ZONE	TYPES OF BUILDING PERMITTED
Residential	Family houses, schools, churches, hospitals
Multifamily	Apartments, duplexes, townhouses, etc.
Commercial	Rental stores, service stations, eating places, small office buildings
Light Industrial	Small plants that produce very little noise, odors, and pollution
Heavy Industrial	Large factories such as steel mills, refineries, chemical plants and the like
Farm	Large tracts of land devoted to raising crops or livestock

Figure 2-2: Common zones for buildings.

Actually, homes have been built in most zones. However, homes built in either a residential or farm zone will have the greatest assurance of not losing their value due to rezoning. Homes built on heavily traveled streets and highways that are now zoned residential may be rezoned commercial later.

When inspecting any home, check with the local building permit office and see if the city or county building inspector or engineer has a zoning map. Then check to see what zone the house you are inspecting is in, and also what zones are nearby.

Today, zoning laws may be so specific in residential areas as to limit house size, placement and even design. Again, zoning information can be obtained from the local building department.

Zoning Variances: If a financial burden will be placed on a property owner unless he or she is allowed to violate a zoning law, the local zoning board may grant a variance that exempts the property owner from that law. Sometimes a variance benefits the community as a whole. For example, a nonpolluting, well-designed, low-profile and highly taxable office building might be an asset to an area zoned for strictly residential use. If too many variances are allowed, though, an area's primary use could change.

Easements

All home inspectors should become familiar with the term *easement*. Easement is a legal term that means a right that is separate from ownership of the land. The most common easements are the rights of utility companies to run telephone or electric power lines above private property or water lines under it. The landowner still owns this piece of property, but he or she cannot put the land to any use that would interfere with the purpose of the easement. In some cases, an easement might be granted to allow another landowner to install a road to his or her property, although this is normally handled in the form of a "right-of-way."

The contour map in Figure 2-3 shows contour lines, feet above sea level, high points, easements for telephone lines, buried cable, and property lines. When contour lines are far apart, as at, land has little slope; the closer together the lines, the steeper the slope. When lines come to a point, this is where water drains. The points aim upstream.

Building Sites and Landscaping

Building Setback

A setback is the distance a building must be set back from property lines, as shown in Figure 2-4. This simple sketch of a piece of property

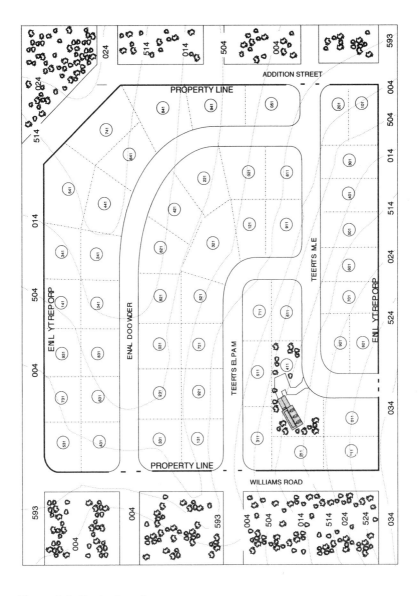

Figure 2-3: Typical contour map.

is called a *plot plan*. Note that the front setback is measured from the front property line, which is 16 feet from the curb, and is shown as 45' B.L. (building line). Sometimes the front building line is measured from the center line (CL) of the street. Side setbacks are usually measured in feet (such as 15' S.B. in Figure 2-4), but sometimes they are measured as a percent of the width of the lot. I.P.S. (iron pipe at side) and I.P.F. (iron pipe at front) refer to pipes that mark corners of the property.

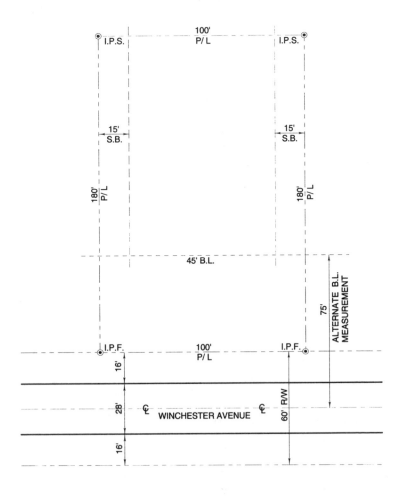

Figure 2-4: Plot plan showing required setbacks.

The local building office will have much of this information that is needed on building setbacks. While there, also try to obtain a copy of any zoning laws.

EVALUATING HOMES

The value of any house is created, maintained, modified and destroyed by the interplay of the following three great forces:

- Social ideals and standards

- Economic adjustments

- Political or governmental regulations

Social forces include population growth and decline, marriage, birth, divorce and death rates, attitudes toward education, recreation, and other instincts and yearnings of people.

Examples of economic forces include natural resources — including location, quantity and quality, industrial and commercial trends, employment trends, wage levels, availability of money and credit, interest rates, price levels, tax loads, and the like.

Political forces include building codes, zoning laws, public health measures, fire regulations, government guaranteed loans, government housing, credit controls, etc. Each and every one of these many social, economic and political factors affects cost, price, and value to some degree. The three of them interweave and each one is in a constant state of change.

Soil Condition

You can learn a lot about the soil at a site simply by looking at the surface. Look at the obvious things first. Big stones or boulders above the surface mean more big stones and boulders under the surface. Bare spots where little or nothing grows may mean a rock ledge just beneath the surface. Heavy clay topsoil means that water will usually run off. Sandy topsoil means that water will be absorbed into the ground. Big trees mean big roots that could eventually do damage to the building's foundation or underground septic tanks or drains fields. Heavy underbrush is a sign that the soil holds moisture well.

If a soil sample is required to complete your inspection, bring a spade with you when you visit the site, using it to dig down about a foot. Topsoil is usually not over eight inches deep, so you don't have to dig far to look at subsoil (Figure 2-5). Subsoil may be quite different at different points on the same site. However, you can generally tell whether you have clay, loam, gravel, or rock to work with. The best subsoil is free from large rocks, drains well, and has good bearing qualities.

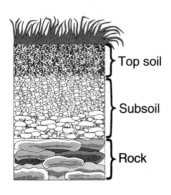

Figure 2-5: Sectional view of the various layers of soil.

The home inspector will not have to make soil tests very often unless the landscaping or sewage system indicate otherwise. For example, bare spots in a yard indicate some type of problem; there is some reason why grass is not growing in this area. Another common problem is much surface water.

One other point to consider here is the level of the water table. A *water table i*s the point below the surface where soil is soaked with water. That point could be down two feet or twenty feet. A high water table will not only cause problems with drainage, but it may interfere with the proper operation of a septic system. Your city or county engineer should have data on water table levels in the areas where you are inspecting.

All building sites that you inspect has contours, as does the house. A contour is an outline of anything. A house will look its best, and will offer more security to the lending institution, if the contours of the site is blended with the contours of the house.

Houses with long horizontal lines look best on sites with uneven contours. Houses with vertical lines look best on nearly level sites. The reason is contrast. The Alps of Europe are beautiful because of their sharp, rugged contours. Just as beautiful are the narrow little valleys that lie between the mountains. It is this contrast between mountains and valleys that brings beauty to both.

When inspecting a home, walk over the building site and see if the house takes advantage of a slope, a stand of trees, a background of hills or woods. At the same time, look for marshy areas, called *swales*. In general, the contours of the site should be an advantage as far as appearance of the house goes, but not a disadvantage in terms of land use.

The Neighborhood

In the mid-1940s — immediately at the conclusion of World War II — when housing was in great demand, many builders developed large tracts, with dozens of houses built from one or two basic floor plans. They all looked alike — same style, same materials, same shape, same window placement, and so on. The only difference between them was color of siding or front door. The houses sold well, because people needed a place to live, but today those neighborhoods look tacky.

Later, smart developers worked with a number of builders, and spaced the lots they built on. Each builder used a half a dozen or more different house designs. As a result, no two houses in any block look alike, and the neighborhood has a pleasant variety that increased property values. Contrast then, makes a lot of difference in the "feel" of a given neighborhood.

When inspecting an individual home have a look at the neighborhood. Are houses and lawns well kept? Are there newer houses in the group? Is there variety? The answer should be "Yes" to all three questions for any given home to be worth its maximum value in comparison to its size.

If there are no other houses close by — say, within 100 yards in any direction from the house you are inspecting — the style is not quite as important. But when houses are close together, the house you are inspecting should:

- Blend with surrounding homes. A two-story colonial rarely looks well among one-story contemporaries.

- Be in the same price range. The house you are inspecting will sell more easily if it is about the same size as its neighbors.

- Fit the neighborhood socially. People like to live among people who are like themselves. A home designed for a young family with small children doesn't belong among homes owned by retired people.

As you inspect each site, also look beyond the immediate neighborhood. In evaluating any home, you must try to decide what the family will be like that will buy the house. Now go one step further. What is that family going to need nearby? Schools, shopping centers, churches, neighbors, jobs? Are these facilities handy? This is very important for one reason. The house that you are inspecting must offer the appropriate life style to be worth its maximum value.

No other single element is more important in the selection of a house than its location. Checking out a location for livability and investment potential may take some time, but it is worth it — to both you and your customers.

LANDSCAPING

While the structure of a house is an important aspect of construction, the owner must also consider the arrangement of the building site: the placement of the house on the lot, driveways, outdoor living areas, trees, shrubs, and hedges. All of these outdoor fixtures fall under *landscaping*.

The landscaping around a house can increase or decrease the value of the house up to 25 percent. Because of this, the home inspector must have a working knowledge of landscaping.

The Building Site

Every building site is unique. The shape, size, kind of soil, and the ground level fluctuations all vary from site to site. The house, outbuildings, driveways, and landscaping should be positioned attractively on the site and placed in areas on the site that accommodate the type of building or structure.

To do an adequate inspection job, the home inspector should walk the site. While doing so, look for any weak points. At this stage of the

Building Sites and Landscaping

inspection, it is more important to note the site's weak points than its strong points to evaluate the property. However, any good features should be mentioned in your report.

There are several things to look for when walking the site: the drainage, the soil condition, the contours, any obstructions, the utilities, the neighborhood, and the value of the land. After observing each of these, answer the following questions:

- Does the land slope so that rain and melting snow drain away from the house?

- Is the house situated or laid out so that the view is best from the living area?

- Are outdoor living areas private?

- If a children's outside play area is included on the site, is it visible from key areas inside the house.

- Is the driveway wide enough; is there ample parking and turning area; does it slope so it can drain water away from the house?

- Can people get out of vehicles and walk to entrances without stumbling over any obstacles?

Drainage

What happens to rainwater or melted snow that falls on the site? Does water drain off or does it remain in pools? Is the house built on a high part of the site? The ideal spot is a knoll or rise from which land slopes away in all directions. See Figure 2-6 on the next page. A gentle slope is also good. Problems can develop when water runs off across the site (Figure 2-7 on the next page) or runs from other areas down toward the site (Figure 2-8 on page 51). The worst condition is one in which the water doesn't drain at all (Figure 2-9 on page 51).

Walk around the edge of the property. What happens to water from surrounding land, especially from downspouts and large paved areas that may drain in the direction of the site? Look for places where topsoil has washed away and subsoil is eroded.

Figure 2-6: One of the best arrangements is a site with a large knoll, allowing the land to slope away from the house in all directions.

Is the area prone to flooding? Ask neighbors and/or the city manager. Check to see if the city has plans to make drainage changes.

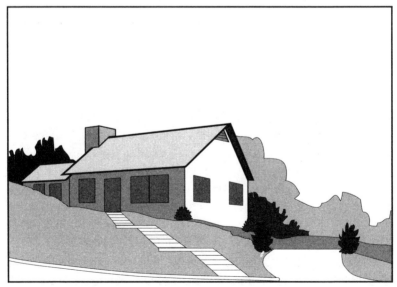

Figure 2-7: Drainage water runs across this site.

Building Sites and Landscaping

Figure 2-8: Water runs from street level to the house on this site.

Drainage problems on any building site can be corrected, but sometimes the correction can be extremely costly. Until the corrections are made, the property value may not be very high. The client should be informed of any serious drainage problems via the inspection report.

To understand how to detect home sites with drainage problems,

Figure 2-9: Perfectly flat building sites that don't allow any water runoff are the worst.

examine the four site plans in Figure 2-10. The site with the best drainage is shown in (A); (B) and (C) offer some drainage problems; (D) is the worst of the four. Visualize these four drawings each time you visit a site and compare observations that are similar to the points in the drawings.

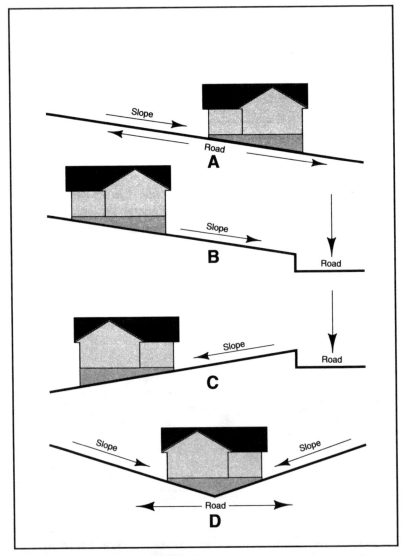

Figure 2-10: Drainage conditions.

Building Sites and Landscaping   53

Utilities

Mark down locations of electrical lines and any gas, water, and sewer lines. Indicate on the site sketch where each of these lines enters the house. Utilities fall under the landscape category. Unsightly power and telephone e lines detract from the overall appearance of any home. If these wires are run near trees, there is always the possibility that the wires will be damaged by falling limbs in a severe storm.

Soil erosion can expose underground cables that could be damaged and can even be highly dangerous to humans and animals. The home inspector should be aware that direct-burial cables (those buried in the ground without any mechanical protection such as conduit) can be damaged by tree roots. Water supply and sewage pipes can also be damaged by tree roots. Basically, landscaping and utilities should be arranged so that they do not interfere with each other.

Contours

In general, the contours of the site should compliment the house, and vice-versa. Look at the group of houses in Figure 2-11. All were built by the same contractor using the same basic design and materials. They all look alike. One way to change the appearance of these homes is through proper landscaping. Each house will look different from the others, even though they use the same design.

Figure 2-11: Though these row houses look almost exactly alike, proper landscaping can change each house's appearance, increasing the property value.

INSPECTING LANDSCAPING

When approaching the site, observe the lay of the land. Is the area on high ground or is it low-lying and damp? Sites which are located on high ground can be expected to drain well. Any ground which slopes down toward a building will be carrying storm water with it. This condition will cause dampness and leaks. If possible, the ground can be reworked to provide drainage away from the building. High-ground sites are almost always more desirable than those of low-lying areas. If a nonpublic septic system is required for the property, the most economical solution is the standard drain field with septic tank and drain tiles. As long as the property "perks" (soil drainage characteristics approved by the local health department), then a drain field running away from the house across the site grade lines is economical.

High-ground sites are not always ideal. Winds can cause terrible living conditions. Cold winter winds can cause great discomfort. Note whether there is a vestibule or enclosed foyer to help partition the house off from weather extremes.

Low-lying sites tend to have storm drainage problems. Note the spacing of street storm drains (usually built adjacent to the sidewalk curbing in the street). If there is an accumulation of dirt, leaves, and trash at the drain, there might be poor drainage. This may be the first indication of site problems. Low areas may be damp even in dry weather. Look for willow trees on the site. Since willow trees need a great amount of water to live, their presence usually indicates high ground water.

Low-lying, marshy areas can lead to problems with insects which thrive in this type of habitat. A screened porch can alleviate this problem.

Trees and shrubbery can enhance a property. An otherwise nondescript property will show great potential if the landscaping is picturesque and cleverly done. Drab windowless building elevations can be improved with the addition of a shrubbery camouflage or grape arbor. At the same time, large trees overhanging a building can fall during storms and cause considerable damage.

At the end of the site visit take photos of the house, carport or garage, and all outbuildings. Remember to personally examine each of these buildings if requested. An inspection checklist appears in Figure 2-12.

Building Sites and Landscaping

		Yes	No
1.	Local zoning classification for this site? _____		
2.	Has a zoning variance been granted this site?		
3.	Is the property encumbered by easements?		
4.	Has a subdivision plat been filed with local authorities?		

5.
6.
	Restrictive setback dimensions are:
	    Front _____
	    Side _____
	    Rear _____		

7.	Does the local contour map show potential for flooding?		
8.	Is there standing water on the site?		
9.	Does the house seem to take advantage of the natural lay of the land? (slope, trees, views)		
10.	Does ground water run away from the house?		
11.	Are the outdoor living areas private?		
12.	Is there ample vehicular area for entering and exiting, and is the line of sight adequate?		
13.	Is street traffic slow or fast?		
14.	Will trucks or noisy vehicles be a problem?		
15.	Is there a business or activity nearby that may control vehicular or pedestrian traffic?		
16.	Is the neighborhood pleasant and show signs of local pride?		
17.	Does this house conform to the neighborhood and seem to fit in?		
18.	Are there many pedestrians about (young/old)?		

Figure 2-12: Inspection checklist for buildings sites and landscaping.

Chapter 3
Foundations

All buildings require a foundation of some type on the ground or on piles, upon which the building is constructed. The foundation may range from wood posts set into the ground to reinforced concrete footings placed in deep excavations. In general, the foundation is designed to support the building structure so that it will not settle, slip, or slide from its original location. And since the foundation rests directly on the ground, the soil must be strong enough to support the weight of the foundation and the house.

A foundation may consist of continuous reinforced concrete footings and foundation walls for exterior and interior bearing walls of the building. Individual footings or piers are provided for free-standing columns. In some instances, wood or concrete piles are driven into the ground and reinforced concrete pile caps constructed upon the pile tops to provide for support of the foundation. A *footing* is a base for a foundation wall or column, just as your foot is the base that supports the weight of your body.

FOOTINGS AND WALLS

All foundation footings for conventional residential building construction are concrete. The walls, however, can be either poured concrete or else built up from either cinder or concrete block. A poured wall produces a better foundation than one built up from block, although block foundations are entirely adequate for most parts of the country. The deciding factor is not personal preference; it's based on construction costs and the requirements of the local building code.

In areas where block is permitted, a poured foundation may always be used, but the opposite may not be true. Where the local building codes permit concrete block foundations, these are the types most often found. Block foundations are almost always more economical than ones that utilize poured concrete. One factor is indisputable: poured concrete foundations are stronger and have a much better chance of being watertight than those made from concrete blocks.

Footings

Building codes require that footings be beneath the frost line. This dimension will vary for different parts of the country. For example, some northern states may require the footings to be a minimum of four feet; in more southerly parts of the country, it may be as little as 18 inches.

The depth or vertical dimension of the actual footing (not the distance beneath the frost line) is usually the same dimension as the thickness of the wall it supports (see Figure 3-1). The width of the footing is twice the thickness of the wall it supports. These are minimum standard dimensions; they are adequate when soil is firm.

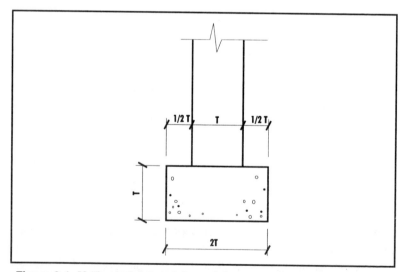

Figure 3-1: Method of determining minimum dimensions for foundation wall footings.

Foundations

Therefore, if the foundation wall is 8-inch thick poured concrete, the footing must be at least 8 inches deep and 16 inches wide, with the wall built in the center of the footing. This leaves 4 inches of footing on each side of the wall.

If the foundation wall is 10 inches thick, the footings must be increased to a depth of 10 inches and a width of 20 inches. In actual practice, however, the width of footings is usually oversized. The width of a back-hoe bucket is usually 24 inches. Therefore, almost all footings for residential construction are dug this width, regardless whether an 8-inch, 10-inch, or 12-inch foundation wall is used. The cost of the additional concrete is more than offset by the savings in labor, if the footings would have to be dug by hand.

The type of concrete mix for most footing applications will be 1:2:4; that is, one part Portland cement, two parts sand, and four parts gravel. Just as important as the concrete mix is the type of soil the footings must rest on. It is absolutely necessary that the bottom of a footing must rest on firm soil. If the soil is not firm, the footings must be placed deeper into the ground until firm soil is reached, or the footings must be strengthened by making them wider, deeper and adding reinforcement rods within the pour. In some areas of the country — especially in coastal areas — firm soil is not reached for great depths. In such areas, piling is normally used, driven deep into the ground by an apparatus appropriately called "pile-driver," which operates much the same as a non-rotary well-drilling apparatus. Several dozen such piles are driven into the ground from 10 to 30 feet, and then the house is constructed onto the piling — the piling acting as the house's foundation.

A footing must also rest on soil that never freezes. The reason for going below the *frost line* — the maximum depth to which soil freezes — is that frozen ground heaves in the spring as it thaws, with enough force to break concrete. Check with your local building inspector to find out what the frost line is in your community. When you come across a house foundation that is severely cracked, chances are the footing was not originally placed deep enough, or it was not placed on firm soil. Correcting such conditions can be a costly affair, and your inspection report should indicate this condition if it is discovered.

All foundations will "settle" to some extent within a year after the original construction. However, cracks in concrete blocks should never be more than a fraction of an inch if the work was done correctly.

In these cases, a minor patching job is all that is required. However, in a poured concrete wall, if a large crack is found, this usually means severe footing/foundation faults. You should call this defect to the attention of your clients so that corrections can be specified in any loan agreements to finance the house and property.

Foundation Walls

As mentioned previously, foundation walls can be either poured concrete or else built up from concrete block. A cross-section of a poured concrete wall is shown in Figure 3-2. This type of wall is constructed by first installing wooden or metal forms. They are then securely braced to hold this position all during the concrete pour. During good weather, the forms may safely be removed within about three or four days. The exterior of the walls and tops of the footings are then waterproofed with an asphalt base foundation waterproofing compound. Two types of waterproofing are generally available, a fairly fluid asphalt mixture that can be brushed on and a mastic type that must be troweled on. The latter is usually the preferred type. An extra heavy coating should be applied at the base of the wall where it meets the footing; the top of the footing should also be covered.

Once the foundation is waterproofed, backfilling is the next line of business. A careful dozer operator can safely backfill a foundation without pushing a wall in, but it is safer to brace the walls from within the basement area. Remember, foundation walls can stand little pressure from the sides until the weight of the house, and its related bracing, rests upon the foundation.

Concrete and cinder blocks are widely used for foundations in many parts of the country where basements are prevalent. They are adequate for most homes, but cannot be considered anywhere equal to poured concrete; block construction is invariably weaker and therefore requires less strain to crack and leak. Poured concrete can also crack, but it can be repaired much easier than a cracked block wall.

Traditionally, block foundations have been more economical than those of poured concrete, but this may not necessarily be the case today. Block foundations require many more worker hours than one that is poured — especially if prefabricated forms are used. Since the wage scale of masons has been steadily increasing, during which period concrete form manufacturers have been developing new labor-

Foundations

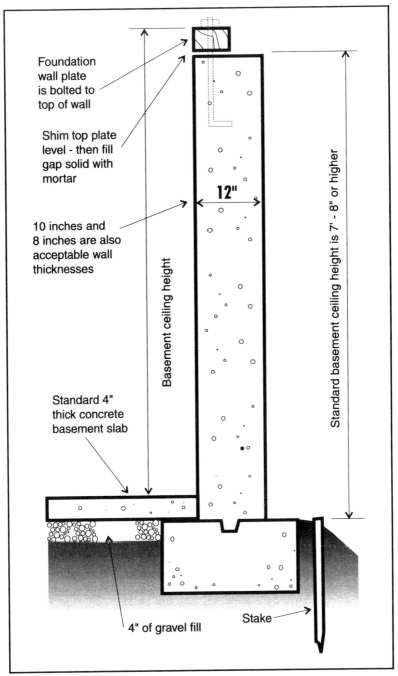

Figure 3-2: Cross-section of poured concrete foundation wall.

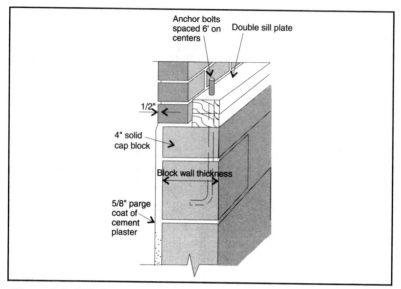

Figure 3-3: Sill plate, anchor bolts, and other details of block foundation construction.

saving forms and techniques of pouring, the poured concrete foundations are once again gaining a small amount of popularity in some areas. But let's take a closer look at foundation walls built up with concrete block. The details of block construction are shown in Figure 3-3.

For standard construction, there are specially shaped blocks to make the job easier. There are "full" and "half" headers that are shaped just right to take floor joists. "Cap" blocks finish off the top of a wall in a very professional and practical manner. For constructing a nicely rounded corner, blocks may be obtained that are already cast with round corners. These are also fine for building columns and posts without having sharp edges. To do the job ordinarily done with heavy timbers over window and door openings, lintel blocks are specified.

Of course, appearance is not a major factor in a foundation wall or in other, similar projects. Here, ease of construction, strength and economy are important. Concrete masonry units meet the requirements under most building codes.

Exterior blocks should be dense and watertight. A variety of block sizes and shapes are available that are suitable for building waterproof walls. Aggregates used to produce the denseness required are gravel, sand, slag, crushed rock and limestone.

Foundations

Piers

A *pier* is a vertical support for a floor or roof structure. There are two types of piers. One type, the standard pier, has a reinforced concrete footing, usually 24 inches square and 12 inches in depth. Four examples of the standard pier are shown in Figure 3-4. Several different materials can be used for piers: concrete block, brick, flat stone, poured concrete, or wood or iron posts. When wood posts are used, they should be treated to resist rot. Piers may also be used as a low-cost substitute for low foundation walls or to support joists in the middle of a house with crawl space.

The other type of pier is a bell-bottom pier. It has no separate footing and can be used only when soil is extremely firm and well packed. Because the hole for bell-bottom piers is wider at the bottom, the wider base of poured concrete serves as a footing. The sides of the hole act as forms for pouring concrete. Bell-bottom piers are used more for vacation homes in northern states than for conventional full-time dwellings. Such piers are also excellent for porch and deck additions. When piers are used in vacation homes and low-cost homes in mild climates, standard piers are usually used.

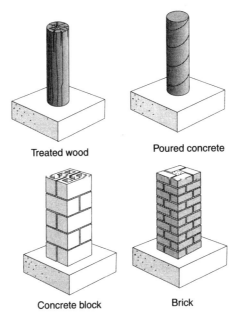

Figure 3-4: Typical standard piers.

Grade Beams

A *grade beam* is exactly what it sounds like — a beam poured right at or near grade level. A grade-beam foundation combines the bell-bottom pier with the continuous wall. Holes for bell-bottom piers are located at regular intervals and then poured with concrete. A continuous trench (earth form) is then dug in line with the piers, and the concrete grade beam is then poured to fill the trench. There are three basic ways to form this continuous grade beam: on top of the ground; in the ground; or half in the ground and half out.

In recent years, treated wooden beams have been used in place of the concrete grade beam. Grade beams are practical only where the ground does not freeze, because the underside of the beam is only a few inches below the surface of the finished grade.

CONCRETE SLABS

Concrete slabs are used extensively in residential construction. For example, the most common type of patio construction is poured concrete. This doesn't mean that you won't find other materials used — bricks, flagstones, soil-cement, or loose aggregates — but poured concrete is the most common. Some are poured smooth, while others will be in the form of grids.

Concrete is durable, fluid enough so that it can be cast in curves, flexible in surface texture and coloring. If the homeowner wants a smooth patio for dancing, a steel trowel finish is called for. A surface to provide traction for footing? The concrete is simply finished with a wood float. Something along decorator lines? Washing with a strong stream of water from a hose just before the concrete sets will expose the aggregates. Flagstone effects may be had with a grooving tool or striated effects with a broom.

From the above, it is easy to see why concrete is the favorite residential patio and porch material. But the entire house can be built (and often is) on a concrete slab. A house built on a concrete slab usually has a short foundation wall set on footings similar to the foundations discussed previously. There are, however, a few additions. First of all, a vapor barrier and gravel base must be provided under the slab. The edges of the slab, however, must rest on top of the foundation wall. Furthermore, the edges should have insulation installed around

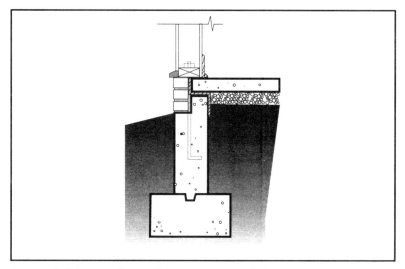

Figure 3-5: Foundation wall for a concrete slab.

the entire perimeter of the slab. Details of slab construction are shown in Figure 3-5. This type of slab is designed to support the weight of the building structure, complete with reinforced concrete footings located below the normal frost line.

A concrete slab used only for patios and porches does not have to have such elaborate footings. Instead, the floor slab is poured with a thickened edge, as shown in Figure 3-6, which serves as a footing and provides the needed bearing surface. This type of concrete floor or slab, however, is not designed to support any building weight; nor is it suitable for basement floors.

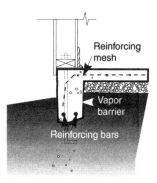

Figure 3-6: Concrete slabs for porches, patios, garages, and the like do not require elaborate footings.

INSPECTING THE FOUNDATION

A basement or crawl space is always the lowest level of a house and parts of it are usually left unfinished by the builder. Even if the owners have a finished recreation room, and perhaps a finished laundry room, there will almost always be some sections left unfinished; that is, the furnace or utility room, storage space, and perhaps a basement garage. If you are inspecting a house that has unfinished areas, the exposed foundation walls, overhead floor joists, and other exposed parts of the house can tell you a lot about the structural soundness of the house. This not only applies to the foundation, but also to the wood framing as well as the electrical, plumbing, heating ventilating and air conditioning systems. Parts of each will usually be exposed in unfinished basement areas.

Crawl spaces offer the same advantages; that is, parts of the floor joists, electrical wiring, plumbing pipes, etc., normally can be seen from the crawl space of any house. Building codes require that the crawl space has at least one entry for repairs, maintenance, or for fighting fires. This is your access point to examine the crawl space of homes you are inspecting. You want to make certain that you have a good flashlight when inspecting such areas. Although the *National Electrical Code®* now requires at least one lighting fixture — with an ON/OFF switch — in every residential crawl space, there is always the chance the lamp will be defective. Furthermore, always carry a pair of coveralls in your car, and also some slip-on rubbers to protect your shoes. A crawl space is just what the name implies; that is, you're going to have to crawl on your hands and knees to move around in the area.

In most cases, a qualified home inspector can get his or her information from the entrance of a building's crawl space — without actually crawling inside. However, in other instances, you will have to actually crawl inside the space to complete your inspection. In doing so, there are certain precautionary measures that you should adhere to. First of all, make sure you wear your hard hat. Due to the limited space, a hard hat can save many bumps and bruises on your head if it should hit the joists above. Be especially cautious of projecting nails, sharp metal hangers (such as pipe hangers) and the like. Remember, these sharp objects don't know the difference between wood and flesh and bones.

Without meaning to sound too alarming, you should also glance around for snakes and spiders — depending upon the geographical location of your inspection. Copperheads and rattlesnakes love the cool environment of crawl spaces during a hot summer day. Snakes are admittedly rare in populous areas, but they have been found in our Nation's Capitol. Naturally, they are more prevalent in rural areas. The chances of actually seeing a snake in a residential crawl space is rare, but it won't hurt to glance around before you enter such areas. Now let's get on with the actual inspection.

When looking in the crawl-space areas, check for signs of termites and other harmful insects that can eventually affect the structural stability of the house. Details of what to look for are fully described in a later chapter. Make a note of the plumbing piping; that is, galvanized, copper, etc. Do you notice any leaks? Check the exposed electrical wiring and any heating ducts. See Figure 3-7. Again, exact details of these items are covered thoroughly in later chapters. They are mentioned here because we are talking about inspecting foundations and crawl spaces, and this is a good place to start looking for harmful insects, structural members, electrical, and related areas of the home.

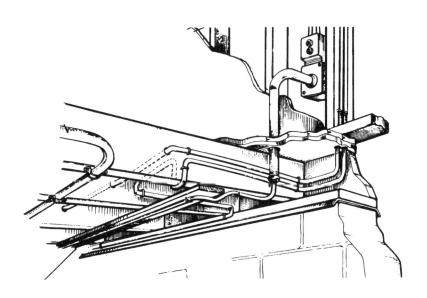

Figure 3-7: Since there are always some unfinished areas in the basement, this is a good place to check on other building systems.

One of the greatest foundation defects that will be found in older homes is the quality of concrete used for footings and foundation walls. Impurities have an adverse effect on any concrete and before the ready-mix concrete trucks, concrete was usually mixed on the job site. Sometimes the aggregate and sand were obtained from local creek beds without "washing." Concrete mixed with these impurities did the job (for the time being), but such concrete deteriorates much faster than concrete mixed with purer ingredients. In this case, look for signs of deterioration such as crumbling or flaking; also large breaks or chips. Moisture, however, can cause the same condition. If this is the major problem, look for stains on the walls and ceiling of the basement or crawl space walls. These are common signs of leaks or seepage. Also look for mildew and wood rot in the basement or crawl space ceiling beams. Does the basement have a musky odor? This is another tell-tale sign of a damp basement.

Wet basements can cause extensive damage, and correction of the problem is often quite expensive. Check the condition of the exterior foundation walls around the house. Are there cracks and signs of water penetration (Figure 3-8)? If the cracks in the foundation wall are near the surface, minor digging is required. Then the crack can be repaired and waterproofed. However, it is sometimes necessary to excavate around the entire perimeter of the building — down to the footing — and lay drain tile along the outer foundation wall to direct the water

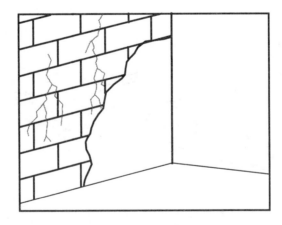

Figure 3-8: Hairline cracks in foundation walls can be detected with a careful examination.

Foundations

away from the house. At the same time, all cracks are repaired and all walls are finally given a coat of waterproofing mastic. The trench is then backfilled. Obviously, such an operation can be quite costly — running into a six-figure amount.

In recent years, several products have come on the market that are advertised for use on the inside of basement walls for waterproofing. Before going to the expense of excavating entirely around the outside perimeter of the basement walls, it may pay the homeowner to try this method first. This treatment, along with a suitably sized dehumidifier, will sometimes do the trick.

When inspecting homes with crawl spaces, always check for a vapor barrier. It is usually a thick polyethylene sheet laid on the earth underneath of the house. As the name implies, a vapor barrier helps to keep moisture from rising into the living area of the house. In slab construction, where the framing is built directly on top of the concrete slab, the vapor barrier is laid directly beneath the slab over the final layer of gravel spread over the ground. The same is true for basement floors.

A checklist for basement and foundation inspections is shown in Figure 3-9 on the next page.

INSPECTION CHECKLIST — FOUNDATION

		Yes	No
1.	Is the house situated on an elevated part of the lot for good drainage?		
2.	Does water flow away from the house (not settle into the foundation)?		
3.	Are the foundation walls free of vertical cracks?		
4.	Are piers in the crawl space or basement free of cracks?		
5.	If cracks exist, are they hairline cracks rather than V-cracks?		
6.	Does the crawl space have adequate ventilation?		
7.	Are the walls straight; that is, no bows or obvious curves?		
8.	Does the house have drain tile around the perimeter of the foundation?		
9.	Are vapor barriers properly installed?		
10.	Does the house smell clean (not musty)?		
11.	Are the basement walls dry?		
12.	Does the slab floor feel dry?		
13.	Are there any large trees close to the house whose roots may damage the foundation or footing?		
14.	Are there adequate floor drains in the basement floor?		
15.	If a sump pump arrangement is used, does the basement floor have the correct taper for adequate drainage?		
16.	Are there any signs of settlement (sunken floor, cracks in walls, floor not level)?		
17.	Are there signs of infestation?		
18.	Does the basement have an outside entry?		

Additional features:_____

Repairs needed:_____

Figure 3-9: Inspector's checklist for house foundations.

Chapter 4
Building Structures

In general, six basic types of building construction are normally found:

- Wood frame
- Masonry
- Reinforced concrete
- Structural steel
- Steel frame
- Prefabricated structures

In many cases, two or more of the basic types of structures are incorporated into a building or sections of it. For example, most residential buildings, regardless of structure type, will have a masonry or reinforced concrete foundation. Then the remaining structure may consist of wood framing, masonry, brick veneer on wood framing or any of the other types of construction.

Wood-Frame Structures: The most common form of building structure for residential applications in the United States is the wood-frame building, with various types of outside finishes — including brick veneer, stone veneer, clapboard, and the like.

Masonry Structures: Probably the simplest form of building structure is masonry, which is erected by placing, one upon the other, small units of clay brick, clay tile, or cement blocks bonded together with cement mortar. However, except for ground-floor or basement slabs, the floor, ceiling, and roof construction of masonry structures is usually of wood frame. Masonry is used also in conjunction with reinforced concrete and structural steel in buildings to form portions of the exterior walls and interior partitions.

Reinforced Concrete Structures: Reinforced concrete construction is commonly used in many sections of the country for multifamily units; that is, apartment houses. Such construction requires the building of forms — either from plywood or prefabricated metal forms. Once the forms are erected, the necessary steel reinforcing rods are inserted within to form the foundations, walls, floor slabs, and the like. Concrete is poured into or onto the forms. When it has hardened or "set up" sufficiently (usually within 28 days), the forms are stripped off. Reinforced concrete is also used in conjunction with structural steel in buildings for the construction of floors, exterior walls, and some interior partitions.

Structural Steel: A framework of steel columns and beams is constructed with sections of various sizes of steel I-beams, angles, channels, and plates, which are usually riveted together. In some instances, particularly in the case of small-sized members, they may be welded together instead of riveted.

Steel-Frame Structures: Open steel-frame construction, usually with exterior walls of masonry or galvanized corrugated sheet steel, is sometimes used in garages and other outbuildings around residences — especially around farm dwellings. The floors are usually a concrete slab on the ground. The roofs are of the tar-and-gravel built-up type on wood sheathing or corrugated sheet steel.

Prefabricated Structures: Prefabricated construction has been used extensively in recent years for residential buildings. Such construction is usually of wood frame with plywood or plasterboard exterior and interior sheathing. Sections of floors, walls, and roofs are constructed at a central yard and the proper sections shipped to the building site, where they are assembled by contractors. Many steel-frame structures are also prefabricated before delivery to the job site.

Building Structures 73

WOOD CONSTRUCTION

Wood has been used for building homes for centuries. In some buildings, the first or ground floor consists of a concrete slab poured on the graded and prepared surface of the ground. In such cases a layer of crushed rock or gravel is usually spread on the graded ground surface, and wire mesh reinforcing is put in place, over which the concrete is poured. See Figure 4-1.

Wood-frame floor construction consists of horizontal floor joists of two-inch-thick or larger timbers on edge and of a width and length required by the length of the span between the supporting walls and the load to be carried by the floor. Such floors are usually rough-sheathed with plywood nailed to the topside of joists, with the finish floor placed over the rough sheathing. Sound-deadening or insulating materials may be placed in the frame floor construction. See Figure 4-2 on the next page.

Walls: The walls of a building serve two purposes: primarily to enclose and subdivide the building into rooms and also to support the roof structure or additional floor and wall structures of multifloor buildings. The type of construction depends upon the purpose served and the weight to be supported.

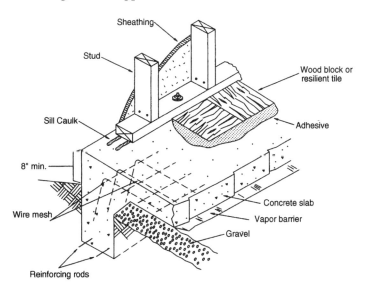

Figure 4-1: Details of a wood structure built on a concrete slab.

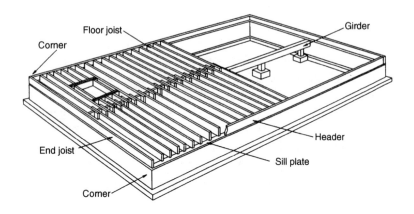

Figure 4-2: Floor framing details.

In the case of frame construction, the walls consist of vertical wood studding to which the exterior sheathing is nailed and the interior wall surface applied. Diagonal wood bracing of one form or another is nailed or otherwise secured between the studs to keep the studs in a vertical position and also to make them less susceptible to twisting. Wood blocking is often secured between the studs to give the walls rigidity and, in the case of enclosed walls, to serve as fire blocking. The studding is usually nailed to a horizontal bottom or sole plate consisting of a two-inch-thick board and to a horizontal top plate or partition cap consisting of two two-inch-thick boards as shown in Figure 4-3.

Roofs: Wood-frame roof construction varies from flat roofs, with only a slight pitch, to relatively steep-pitch hip or "A-frame" roofs. Special designs of "sawtooth" construction and self-supporting wood-frame arch construction are adapted to some forms of construction, but in the majority of residential buildings, wood-truss supporting construction with wood purlins is covered with plywood sheathing. Typical roof framing is shown in Figure 4-4.

Types Of Wood Structures

Wood structures are built in many different ways, ranging from pole construction to sophisticated, complex designs. Since wood structures

Building Structures

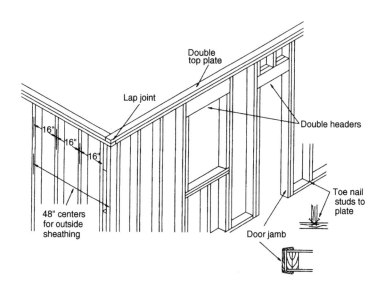

Figure 4-3: Principles of framing for door and window openings.

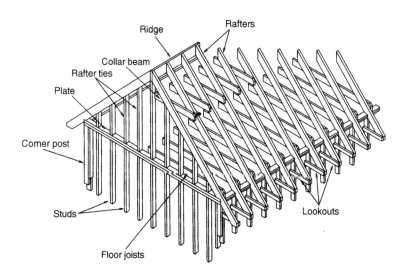

Figure 4-4: Typical roof framing.

are used far greater than any other type of residential building construction, let's take a look at some of the more popular designs.

Post and Beam Construction: This type of building system uses heavy posts and beams to form the base framework. The use of timbers creates a load-bearing system that allows an open arrangement. The wall areas, except for the posts, is nonbearing. Heavy planks are used on the floor and roof. Widely used for ultra-modern residential construction where cathedral ceilings are prevalent.

Pole Construction: This type of construction is popular for storage buildings and also for farm buildings. Poles are used as the main support members.

A-Frame Construction: An A-frame construction process is often used for summer or vacation homes, especially in the eastern United States. As the name implies, the house is constructed with an "A" shape when viewed from the ends.

Balloon Framing: This type of house framing was once popular during the early part of this century. In this type of construction, the studs are continuous from the foundation sill to the roof line as shown in Figure 4-5. Joists at each floor are nailed to the side of the studs and rest on a ribbon or ledger board.

Platform Framing: This is a widely used framing method for residential construction that utilizes a complete platform at each floor level. Walls are constructed separately and are then raised in place on the platform. Refer back to Figure 4-2.

MASONRY STRUCTURES

Masonry construction of all types has proven its usefulness for centuries. Bricks, concrete blocks, tiles, and other variations have been used for all types of construction from gate posts to huge office buildings.

For economy in most types of commercial construction, the standard 8 x 8 x 16-inch and the 4 x 8 x 16-inch concrete blocks still dominate the field, but many new shapes and sizes are also available. Some, too, are colored, or they have polished or cut face to resemble stone. Large and small sizes and shapes are also used to vary the pattern in walls. Lintel and jamb blocks for doors and windows are prefabricated, ready for use.

Building Structures

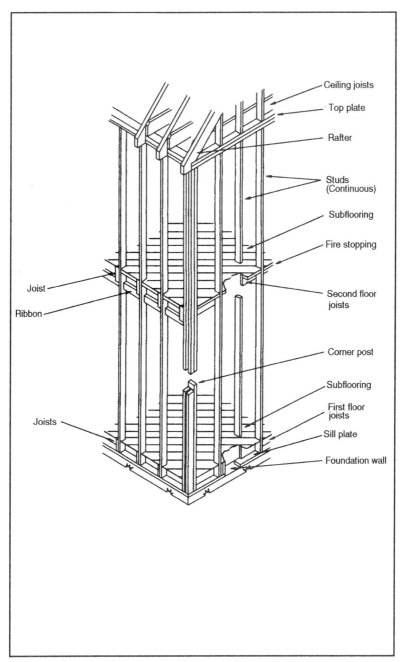

Figure 4-5: Balloon framing was very popular for residential construction during the early part of this century.

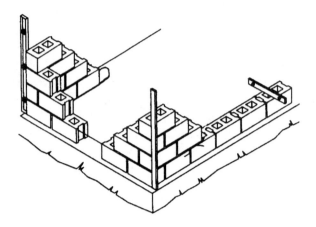

Figure 4-6: Details of concrete block layment.

Concrete blocks are used mostly for foundation work in residential construction, or to provide firewall construction in town houses or multifamily developments. See Figure 4-6. For interior basement walls, furring strips are often used over the blocks, and then some type of wallboard is used for the interior finish.

Standard bricks are frequently used to face concrete block when the block is above ground. Wood structures are often brick veneered and usually provide better insulation than when brick is used on concrete block walls. See Figure 4-7 .

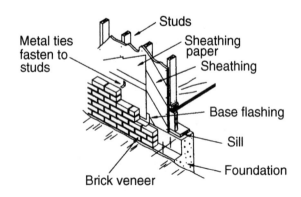

Figure 4-7: Brick veneer.

Building Structures

Regardless of the type of masonry used, all are bonded together with mortar. To be good for bonding together concrete blocks or bricks, mortar must have sufficient strength and be workable. Unless properly mixed and applied, the mortar will be the weakest part of any type of masonry. Both the strength and the waterproofing of masonry work depend largely on the strength of the mortar bond. Consequently, the home inspector must always check the mortar joints of masonry construction. Weak joints cause water leaks that will eventually weaken structures, and may jeopardize the lending institution's investment.

Bricks and Brickwork

Standard bricks manufactured in brick factories usually are $2\frac{1}{4}$ x $2\frac{1}{4}$ x 8 inches. Handmade bricks, along with English and Roman bricks, are sometimes of other dimensions. Figure 4-8 gives brick terminology.

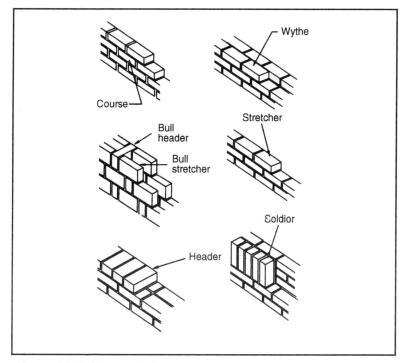

Figure 4-8: Brick terminology.

Brick Veneer: In brick veneer construction the brick is used only as a facing material, without utilizing its load bearing properties. The brick is applied over wood framing and sheathing of both old and new houses.

Veneered walls resist fire exposure better than frame walls. The brick can be laid in any type bond or pattern formed by the use of half-bricks as headers.

In new construction, the foundation walls should extend 5 inches outside the face of the wood sheathing to receive the brick veneer. On old buildings, the veneer should be started on the projecting portion of the footing.

The wood walls should be covered with waterproof building paper. The brick veneer is then anchored in place with one noncorroding metal tie for each two square feet of brick area. The ties should be spaced not more than 24 inches apart horizontally and vertically.

Brick veneer can also be applied on concrete walls or more commonly, cement block walls. A shelf angle secured by an anchor bolt imbedded in the concrete supports the brick veneer or brick ties are embedded in the mortar joints of concrete block walls. Dovetail anchor slots are provided to hold the brick to the wall. The space between the brick and the concrete or concrete block wall is slushed with mortar or grout.

Concrete Block Walls

Because of their strength and fireproofing qualities, concrete blocks are one of the most widely used materials in masonry work.

Blocks are available in colors that are durable and easily maintained. Concrete blocks can also be spray painted. A waterbase paint such as colored Portland cement or a paint with a latex-vinyl or acrylic-emulsion base may be used. Before painting the wall, a sealer coat is applied to close up all pores in the cement.

Prefabricated Structures

Prefabricated homes have become quite common over the past couple of decades. Sometimes called *Industrialized buildings,* they use a high degree of prefabricated construction or building components or both. For example, a house may use partially prefabricated compo-

Building Structures

nents such as windows, doors, trusses, and the like. In some cases, a whole section may be built at the factory and delivered to the job site. In many cases, whole houses — usually delivered in two sections — are built at the factory and then joined at the job site.

INSPECTING BUILDING STRUCTURES

A home's structure is just as important as the building's foundation. Defects in either can result in serious problems that greatly affect the overall value of the home.

A broken window pane or door is minor. Both can be repaired or replaced with relatively little cost. However, when it comes to the building's structural framing, it is an altogether different situation. Framing corrections can run into tens of thousands of dollars and can make a tremendous difference in the value of a home.

Since most framing is hidden by sheathing or other types of finishes, most structural framing in existing homes is not readily visible to the home inspector. However, there are still many ways in which this part of the home may be inspected. Once you know what to look for and gain experience in these areas, you should have little difficulty in detecting any structural deficiencies and filing your reports accordingly.

Inspecting Floors

You can easily detect if the home's floors are level using a conventional carpenter's level. Expect all floors in homes that have been built more than five years to be slightly off (due to natural settlement of the foundation), but if the bubble in the level indicator is outside of and not touching the hair marks, you might have structural problems. See Figure 4-9.

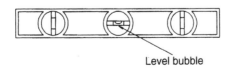

Level bubble

Figure 4-9: A conventional carpenter's level can be used to check the level of floors in the houses you inspect.

Usually, when an entire floor is not level and sags in one direction, it could mean settlement of the foundation. However, if the floors sag inward, toward the girders, there is definitely a problem. That is, if the floors on both the right side of the house and also on the left side of the house sag inward, something is wrong. Seldom will a foundation sink in the middle of a building.

If you find this condition, check the floor joists from below (if possible) for any wood rot or termite infection. Look for cracks in structural members. If this condition is due to settlement of the piers, it can be corrected by using floor jacks to bring the center girders back to their correct height. However, if structural members show weakness or damage, these defective members will have to be replaced to correct the situation. This means that the adjacent structural members will have to be temporarily supported until the replacement has been made.

Squeaks or other noises heard as you walk across a floor can mean that structural repairs are in order. Again, check in the basement or crawl space for damaged floor joists or girders. These noises could mean that the bridging is loose or some of the subflooring or finished floor boards are loose. Correction of the latter can be made by driving thin wedges between the floor joists and the floor areas causing the problem.

In some cases, the floor joists may not be bearing an equal load. If this condition is expected, one possible solution is to try to tighten the bridging on the floor joists. If this doesn't correct the problem, straight pieces of 2 x 8 lumber can be nailed between the joists as shown in Figure 4-10.

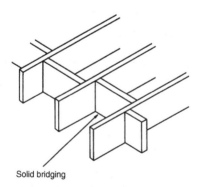

Solid bridging

Figure 4-10: When structural defects indicate that the cross bridging is the culprit, the condition can sometimes be corrected by installing straight 2 x 8 members between the joists.

Feeling undue vibrations as you walk across a floor is an indication that the floor joists are not adequately supporting the subflooring. There are several reasons for this: the floor joists may be spaced too far apart; their span might be too long, or again, a few joists may be carrying all the load due to improper bridging.

One way to test for vibration is to place your weight on one foot, and then bend your knee up and down as if jumping, keeping the foot that you are standing on flat on the floor. This method is especially useful if there is a china closet with cups and dishes in the room. When the structure is adequate, you might hear the dishes rattle slightly. If they rattle excessively, then correction to the floor support is necessary. This correction could entail correcting the bridging, the floor joists, the girders, or the piers supporting the girders. A closer examination will be necessary to determine which.

The most serious conditions are those that involve structural damage caused by either wood rot, termites, or actual cracked or broken structural members. Perhaps the foundation has settled and is not offering the required support to the house framing. One way to check this latter condition is to inspect all supporting surfaces. If gaps are visible, the foundation may have sunk. Wedges or other devices must be used to fill the gap. Sometimes it is necessary to jack up the framing with building jacks and then build up the foundation itself. Obviously, rebuilding the foundation can be costly. If a lending institution is making a loan on the property, you should alert the firm of this condition so the property value for the loan can be evaluated accurately.

Checking Doors and Windows

In a structurally sound house, all doors and windows should open and close easily. If they do not, it could be caused by the swelling of the wood from humidity. However, in most cases, it is due to either the settlement of the foundation or defects in the home's structural framing. If the owners of the home are available for questioning, try to determine if a window or door has been closing hard since the house was built or if this condition just happened recently. If the former is the case, chances are the door or window was not installed properly in the first place. However, if the condition started recently, then there could be some structural problems with the home.

If nearly all windows and doors open and close hard, then a more serious problem exists, which should be noted on your report to the firm or person hiring your services. In this case, you will probably be called upon to find out the exact extent of the structural defects and to determine the estimated costs of having the defect repaired. You can probably give a "ball-park" figure yourself. If you find severe damage, you should recommend that the owners contact an experienced contractor to make the repairs.

Ceiling Inspection

Let's assume that your floor inspection has passed with flying colors. You are about to grade the home that you're inspecting with a perfect score. However, you notice that one of the ceilings in a room is sagging.

A sagging ceiling, like the one in Figure 4-11, or one that is not level, can be caused by a number of things. All of them will be explained thoroughly in the chapter on ceilings, but let's look at some of the structural reasons for this condition now.

You will probably find a sagging ceiling in a home that has been renovated. Many of the older homes had very small rooms when compared with the size of rooms in modern homes. Many older homes were usually heated exclusively with wood-burning stoves or fire-

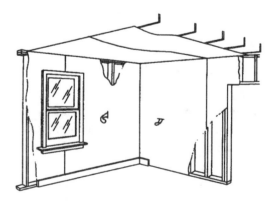

Figure 4-11: Sagging ceilings can be only minor in some cases, but such conditions could indicate a structural problem.

Building Structures

places. The smaller the room, the less amount of heat required. In fact, in many older farm and ranch houses throughout the country, the occupants normally used only the kitchen and dining room for "living areas" during winter months. The remaining rooms (except for the bedrooms) went unused until spring. The wood-burning range in the kitchen, and perhaps a small wood-burning stove or fireplace in the dining room, provided sufficient heat without waste.

However, when these old homes were renovated, these small rooms were made larger to accommodate more guests and to comply with modern trends. To renovate, supporting wall partitions were knocked out. In many cases, bearing beams were not installed to compensate for the stud wall. This accounts for sagging ceilings in many of the older homes throughout the country.

To correct this situation install a larger supporting beam— its size depending upon the length of the span. In some cases, it could require three or four 2 x 12s to carry the load of the rooms above. In either case, considerable expense could be involved, which should be noted on your inspection report.

In other cases, the sagging ceiling may be due to inefficient fasteners for the ceiling finish; that is, plasterboard, drywall, etc. If this is the case, the condition is not too severe. This may be corrected with relatively little expense by going over the ceiling panels and securing them better, and finally, patching the holes with spackling compound, sanding the repairs flush with the adjacent ceiling, and then painting.

A sagging ceiling could mean structural damage to the ceiling joists or related supporting members. If so, you should be able to detect looking at them in the attic. However, if the condition exists on the first-floor ceiling in a two-story house, the condition is not as easily detected.

The checklist in Figure 4-12 should prove useful during structural inspections.

	INSPECTION CHECKLIST — STRUCTURAL	Yes	No
1.	Are the floors level?	___	___
2.	Are wood beams and surfaces free of termites or wood decay?	___	___
3.	Do all windows fit and operate easily?	___	___
4.	Do all doors fit and operate easily?	___	___
5.	Are ceilings level?	___	___
6.	Is the height of each step (tread) the same size?	___	___
7.	Are the walls straight; that is, no bows or obvious curves?	___	___
8.	Is the roof pitch steep enough?	___	___
9.	Are floor joists adequately bridged?	___	___
10.	Have any existing partitions been removed?	___	___
11.	If the answer to No. 10 is yes, were provisions made for adequate support of the framing members above?	___	___
12.	Are there any cracks in the floor, ceiling or walls?	___	___
13.	Are there any cracked or split joists, truss, or rafters?	___	___
14.	Are there adequate floor drains in the basement floor?	___	___

Circle the appropriate answers below:						
Type of roof:	Gable	Hip	Gambrel	Flat	Shed	A-frame
Other:_____						

Overall condition of framing	Good	Fair	Poor
Defects:			

Figure 4-12: Inspection checklist for building structures.

Chapter 5
ROOFS

A roof is the top exterior covering of a house and is used to help make homes waterproof and weather tight. In general, the roof consists of a supporting structure composed of rafters and/or trusses as discussed in Chapter 4 — Building Structures, some type of sheathing on top of the structure, and a roof covering. Besides these three basic items, the home inspector should also be concerned with the deck, eves, soffits, various underlayments, flashing, along with insulation and ventilation in the attic space under the roof.

A home's roof is one of the most essential parts of the home. In fact, it is second in importance only to the foundation and basement. Consequently, the home inspector should become thoroughly familiar with the various roof designs, their names, and how each is constructed. The various types of roof covering comes next. Once you know how a certain type of roof covering should be installed, your job of inspecting the roof will be much easier.

After completing this chapter, you should be able to recognize the various types of roof styles to enable you to accurately describe them in your report. Furthermore, you will be able to recognize the various types of roof coverings and to determine if the covering suits the roof pitch and style of home. Finally, you will be taken step-by-step through the complete inspection process of residential roofs. The checklist at the end of this chapter will facilitate your doing your job in a professional manner.

ROOF STYLES AND DESIGN

The roof line of a single- or multifamily dwelling should be one of your first observations as you approach a home for inspection. It should also be one of the first items to write down (or check) on your on-site inspection report.

Many types of roofs are used in modern building construction. The following are the most common:

- *Shed or Lean-to Roof:* This, the simplest type of roof normally used, has only a single slope as indicated in Figure 5-1.

- *Gable or Pitch Roof:* This is the type most commonly used for residential buildings. It has two slopes meeting at the center or ridge and forming a gable. See Figure 5-2.

- *Hip Roof:* This roof consists of four sides, all sloping toward the center of the building. The rafters run up diagonally to meet the ridge, into which the other rafters are framed as shown in Figure 5-3.

- *Hip and Valley Roof:* This type of roof is a combination of two hip roofs intersecting each other, usually at right angles. The valley is the place of meeting of two slopes of the roof, running in different directions. There is a great variety of modifications of this roof type. See Figure 5-4.

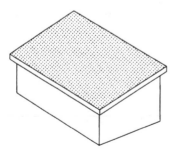
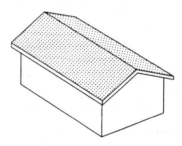

Figure 5-1: Shed (lean-to) roof. **Figure 5-2: Gable or pitch roof.**

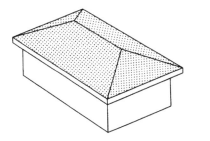

Figure 5-3: Hip roof.

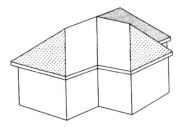

Figure 5-4: Hip and valley roof.

- *Gable and Valley Roof:* This roof style is practically the same as the hip and valley roof except that two gable roofs intersect rather than two hip roofs. See Figure 5-5.

- *Flat Roof:* A type of roof that is almost flat but still has a slight slope for water runoff. See Figure 5-6.

- *Mansard Roof:* This is a four-sided roof with a double slope on each side. See Figure 5-7 on the nex page.

- *Gambrel Roof:* A roof with two slopes on each side, the bottom slope has a sharper angle. See Figure 5-8 on the next page.

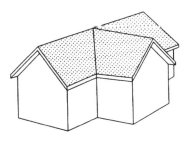
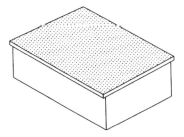

Figure 5-5: Gable and valley roof.   **Figure 5-6: Flat roof.**

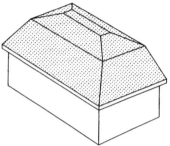
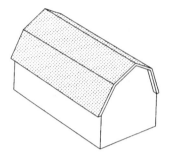

Figure 5-7: Mansard roof.   **Figure 5-8: Gambrel or barn roof.**

Some houses incorporate several of the above styles into one house, making it difficult to assign a single name. If in doubt, use geometric terms when describing the shape of a house roof. For example, you will sometimes run into unusual house designs. Some will be round, others will be geodetic in shape, while others may have a roof that is currently called an "A frame" or "modified A frame." However, as long as you know the basic designs shown in Figures 5-1 through 5-8, and a few geometric terms, you should have the required knowledge to describe any house roof you may come in contact with.

Roofing Terms

Since the type of roof covering depends a lot on the type of roof and the way it is constructed, a review of roofing terms shown in Figure 5-9 is in order.

Span: The span of a roof is the distance over the wall plates.

Run: The run of a roof is the shortest horizontal distance measured from a plumb line through the center of the ridge to the outer edge of the plate. In equally pitched roofs, the run is always equal to half of the span or generally half the width of the building.

Rise: The rise of a roof is the distance from the top of the ridge and of the rafter to the level of the foot. In figuring rafters, the rise is considered as the vertical distance from the top of the wall plate to the upper end of the measuring line.

Roofs

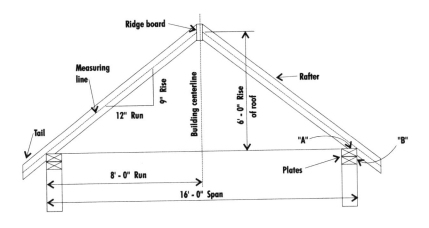

Figure 5-9: Span, run, rise, pitch, and measuring line of a roof.

To find the rise of a roof, multiply the pitch by the span, or

$$\text{rise} = \text{pitch} \times \text{span}$$

Pitch: The pitch of a roof is the slant or the slope from the ridge to the plate. It may be expressed in two ways:

- The pitch is the ratio of the total rise of the roof to the total width of the building.

- The pitch of a roof is also expressed in a number of inches rise to each foot of horizontal run.

The diagram in Figure 5-10 shows the principal roof pitches. They are sometimes called 1/2 pitch, 1/3 pitch, 1/4 pitch, etc. because the height from the level of the wall plate to the ridge of the roof is, say, one-half, one-third, or one-quarter of the total width of the building. The pitch may also be expressed in the number of inches rise to each foot of horizontal run. Therefore, if the rise is three inches in one foot of span, the pitch is 3 to 12; if 4 inches of rise, the pitch is 4 to 12; if 6 inches rise, the pitch is 6 to 12, etc.

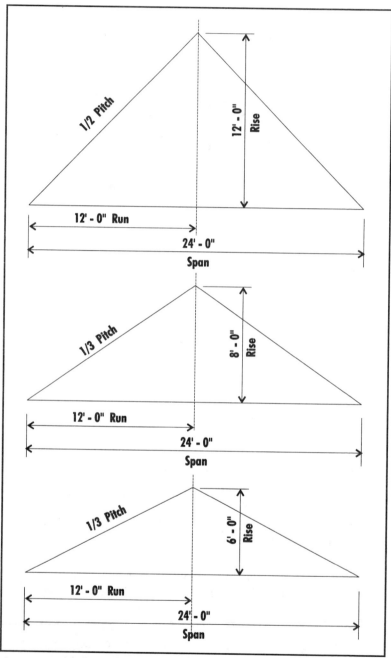

Figure 5-10: Principal roof pitches.

ROOFING MATERIAL

Climate plays a very important role in selecting the type of roofing for a home. In California, for example, where they are seasonal brush fires, it is advisable to use a fireproof roof covering such as cement tile, Spanish tile, or aluminum shake shingles.

In Miami, on the other hand, there is no snow, sleet, or ice. There is a short rainy season, which includes a hurricane season with winds up to 90 mph, plus long periods of hot clear days with intense sunlight. Roofing materials must therefore be carefully selected to satisfy these conditions.

Bangor, Maine, which has rain, sleet, heavy snowfalls, and winds as high as 60 mph, and only a month of bright intense sunlight, roofing materials must be selected to satisfy these conditions. Therefore, the home inspector should be familiar with the year-round climates in any area he or she is inspecting, and make sure the building roofs meet with these conditions.

Wood Shingles

Wood shingles are used in many parts of the United States, but, unfortunately, one of the notable physical properties of wood is that it burns. To give wood shingles better protection against fire, they are treated against this hazard with a fire-retardant material. Wood shingles are class A, B, C, with the A class having the highest fire-resistant rating.

Shingles and other coverings are manufactured as single units or in units simulating shingles and are from 3 feet to 6 feet long. The shingles and coverings are made from materials such as wood, asphalt, aluminum or copper, slate, roll roofing, built-up roofing, corrugated galvanized iron, metal, and tile. On older homes, you will still find asbestos-cement shingles, but the asbestos/cancer scare in the last quarter of this century has made this type of shingle obsolete for modern buildings.

Wood shingles are single shingles manufactured in three basic types: dimension, random, and hand split. Wood shingles for houses should be grade 1, which are all heartwood, all edgegrain, and tapered. For secondary type buildings, such as an outside storage building, boat house, etc., grade 2 shingles are permitted. The wood used in both grades are western red cedar and redwood, and their heartwood (used in grade 1 shingles) has high decay resistance and a relatively slow shrinkage factor.

The butts of first course wood shingles should project $1\frac{1}{2}$ inches beyond the first sheathing board, and should be doubled at all eaves. Spacing between shingles should be $\frac{1}{4}$ inch and their joints in any one course should be separated by not less than $1\frac{1}{2}$ inches from joints in adjacent courses. Joints in alternate courses should not be in direct alignment.

Asphalt Shingles

Conventional asphalt shingles are made of cardboard that is covered with asphalt in which a layer of ceramic granules is imbedded. This type of shingle is relatively long lasting but provides only minimal protection against fire.

The asphalt shingle with a fiberglass base looks like other asphalt shingles, but the core is fiberglass instead of cardboard. This type of shingle provides tremendous fire resistance.

Strip Shingles: Individual asphalt strip shingles are installed on roofs having pitches of 3 to 12 or greater, and must be nailed to solid sheathing. To overcome high winds, asphalt shingles can be obtained with special tabs of adhesives near the butt end of the shingle. Single asphalt shingles are manufactured in hexagon and Dutch-lap type and are available in various colors and weights. Square-butt shingles come in a weight of 235 pounds per square while strip and lock-type are available in a weight of 300 pounds per square. Asphalt shingles also come in weights of 250 pounds per square or more.

The strip shingle is 12 inches by 36 inches, has three tabs, and is usually laid 5 inches to the weather. One bundle of shingles contains 27 strips, and three bundles cover one square of roof surface.

Square-Tab Strip Shingles: Square-tab strip shingles are intended for roofs having a slope lower than 4 to 12 but no less than 2 to 12. The shingles are laid at a 5-inch exposure, where each shingle requires four $1\frac{1}{2}$-inch nails. All tabs are cemented down. A starter strip of 90 pound mineral-surface sheet is applied before the first shingle course is nailed down. In place of the mineral-surfaced sheet, reversed shingles can be applied. Felt plies of underlayment cemented together are first applied on a tight wood deck or roof sheathing.

Square-Butt Asphalt Shingles: Square-butt asphalt shingles are applied similarly to square-tab shingles. Solid wood roof sheathing is covered with an underlayment having a 4-inch endlap and a 2-inch headlap. An eave flashing strip is applied next. The first course of shingles at the eave should be inverted, or else a 9-inch starter strip should be used in place of the inverted shingles. At the rake of the roof the first course should be started with a full strip, while the second course is started with a full strip minus $1/2$-tab. The third course is started with a full strip minus its first tab.

Aluminum Shakes

Aluminum, imitation cedar panels are 4 feet long by 12 inches wide, and are packed in cartons (25 pieces per carton) that cover one square (100 ft.). Each panel is fastened down with three nails, and in areas where high winds are prevalent, four screw shank nails that penetrate to a minimum of $1\frac{1}{2}$ inches into the roof sheathing are recommended. The nailing is done along the nailing flange of the panel.

Before the aluminum panels are applied, a new or recovered roof should be covered with a number 30 felt.

The shake panel is installed from left to right, beginning at the lower lefthand corner of each area to be covered. Begin with any size panel that has a factory formed right end. Lock the first panel into the starter strip, making sure it is completely seated and fitted into the cap or channel on the left side. Secure each panel with three equally spaced screw shank nails, driven into solid sheathing. Nail the left end of panel first, then the right end, and then the middle.

Clay-Tile Shingles

Clay-tile roofing shingles should be used only on roofs having a pitch of $4\frac{1}{2}$ to 12 or greater. The clay used in the manufacture of clay-tiles is the same as that used in the manufacture of brick. The tiles are made either by the soft-mud process or the dry-press process.

The soft-mud process requires a clay-water mixture of 20% to 30% water. This allows the clay to be easily molded. In the dry-press process the clays are mixed with a smaller amount of water (7 to 10%). This keeps the clay almost in a dry state. The clays are then molded under high pressure into specific shapes.

Clay-tiles made by the soft-mud process are less dense and somewhat lighter in weight than those made by the dry-press process.

Clay-tiles are manufactured in various types, such as Spanish Mission, French, English, Norman, and Scandia. They are generally reddish in color, but are available in a variety of colors and color glazes. Clay-tile roofing shingles weigh from 800 to 900 pounds per square foot.

California Mission Tile: California Mission tile, or Spanish tile, has been used on houses and other buildings for centuries. Because of the weight of this clay or concrete tile, larger size rafters with narrower on-center spacings are required. This and breakage are factors that contribute to the overall high installation costs.

In place of the heavy clay or cement tile, a lightweight tile, also known as California Mission tile, is available. It consists of 0.020-gauge aluminum or 30-gauge G-90 galvanized steel with a baked enamel finish similar in color to that of the clay tile.

The tile utilizes a two-way interlocking (horizontal) and a two-way overlapping (vertical) installation system that prevents leakage when exposed to rain and very high winds.

Each tile is $7\frac{1}{2}$ inches wide by 17 inches long and has a somewhat wider flange at the top when used as the pan tile. The tile used as the cap tile is reversed so that the narrow flange faces down.

A box of caps and a box of pans will cover a roof area of 100 square feet or one square. This type of tile is installed over 2 inch x 3 inch wood stripping and is used on roof slopes of not less than 3 to 12. The underlayment is one layer of 20-pound asphalt-saturated rag felt over a plywood base or substrate. Galvanized steel or aluminum nails are used.

For a gable roof it is necessary to measure 4 inches from the side of the roof to place the first 2 x 3 furring strip. All other furring strips are then toe-nailed, $10\frac{1}{2}$ inches apart.

The bottom pan with wide flange up, is centered between furring strips. The top edge is set to previously strung guide lines. Nail the pan to the roof by driving a nail 1 inch from the top edge of the pan.

Spanish Tile: This tile is similar to California Mission tile, in that it is metal with a baked-on tile-color finish. Instead of a plain curved metal pan, it is S-shaped, with a nailing flange on one side and an edge-lock on the other.

Before this tile is placed on the roof, an underlayment of 30-pound asphalt-saturated roofing felt is applied, with an overlap of 6 inches on each hip, and a lap of 18 inches on each succeeding width. Underlayment should overlap 1 inch on all valley metal. Spot nail to hold felt in place.

For hips and ridges, 2 x 6 inch wood furring strips are used. These are the same nailing bases as those used for hip and ridge pans of California Mission tile.

Espana Mission Tile: This tile is made of concrete. Oxides and cement are added to the tile surface and sealed prior to placement in a controlled curing chamber. The tile is barrel-shaped and is 3 inches above the surface plane.

Slate Roofing

Slate shingles are heavy materials, weighing 700 to 810 pounds per 100 square feet. The shingle is 20 to 24 inches long in increments of 2 inches. The surface of the tile may have either a smooth or a rough finish. The width of a slate shingle is 6, 7, 8, 9, 10, 12 or 14 inches. Two holes for nailing are pre-drilled at the top of the shingle. The thickness of the slate shingle varies from $\frac{1}{4}$ to 2 inches in $\frac{1}{4}$-inch increments.

Asphalt Roll Roofing

This roofing material is used for low roof pitches of 1 to 12 and up to 4 to 12. It is manufactured in rolls 36 inches wide and 36 to 48 feet long. Of the 36-inch width, 17 inches — the exposed part — is usually covered with a colored mineral material. The remaining 19-inch wide plain surface is overlapped by the next roll.

Built-Up Roofing

Built-up roofing is usually designated as 3-ply, 5-ply, etc. It consists of alternate layers of asphalt-or tar-saturated felts and hot or cold asphalt or hot tar (known as pitch). Each layer of material except the first is called a ply. The first layer is not considered a ply because it consists of one or two layers of unsaturated felt, nailed down to the substrate. This layer is important because it is used to bond the built-up roofing to the substrate.

A 4-ply built-up roof, for example, consists of one ply of unsaturated felt, and four plies of saturated felt and pitch. The top ply is covered with slag or stone chips to protect the surface from intense sunlight.

Built-up roofing is applied mostly to flat roofs and to roofs with a pitch that does not exceed 2 to 12.

Corrugated Roofing

Corrugated roofing can be used for roofs having pitches of 3 to 12 and greater. The corrugated sheets can be of plaster, metal, or wire-glass, and may be joined at the sides by overlapping the corrugations. End laps are generally 6 inches. All sheets except the plastic type require pre-drilling of nailing holes, and all methods of attachment should allow for slight movement caused by expansion and contraction.

The maximum span between supports must not exceed 2 feet 6 inches for lightweight sheets and 4 feet 6 inches for standard weight sheets. Corrugated asbestos-cement sheets are available in lengths of 4, 5, 6, 8, 10 and 12 feet.

Fiberglass-reinforced corrugated plastic sheets are available in widths of 2 feet $3\frac{1}{2}$ inches, 2 feet 5 inches, and 4 feet $3\frac{1}{2}$ inches and in lengths of 4, 5, 6, 8, 10, and 12 feet. The maximum allowable span between supports is 4 feet.

Metal Roofing

Metal roofing is used on roofs having pitches of 3 to 12 and greater. Metals such as lead, stainless steel, aluminum, copper, and terneplate have been used in the past.

Terneplate or terne-coated stainless steel (TCS) is covered on both sides with terne alloy (80% lead and 20% tin). When properly installed it can be maintenance-free for many years. It should outlast the building on which it is applied. Terneplate is recommended for use in severely corrosive environments.

When copper is used as a roofing material, care should be taken that water running from the roof does not come in direct contact with other materials such as siding, because copper, when exposed to the weather, develops a green coating that can run off and stain other materials.

Structural Metal Batten System

This metal roofing material is 24- or 22-gauge steel with a factory applied weather-resistant coating.

Roofing panels are fastened directly to the structural steel girts with No. 12 to 14 hexagonal washers, $2\frac{3}{4}$ inches long, at each purlin. When this type of roofing is applied over plywood sheathing, self-drilling No. 12 screws are used.

The panel is 18 or 24 inches wide. the bottom heights are $1\frac{5}{8}$ inches of $1\frac{3}{4}$ inches and the panel lengths are 4, 5, or 6 feet. The galvanized batten hold-down clips are of 24-gauge metal, and are attached to side joints of the roofing panel at 24 inches on center.

Other roof panels are roll-formed from 0.032 inch aluminum alloy, or galvanized steel with a weathering copper coating to a thickness of 2-mils and are oven-cured at 650°F.

Fiberglass Roofing

Teflon-coated fiberglass fabric is used as a roof covering over shopping centers and shopping malls. The roof is supported by metal arches running the entire length of the mall. The fabric allows some daylight to pass through, thus conserving energy.

The Teflon-coated fiberglass roof reflects about 80% of the sun's rays while allowing light to enter. This type of roof is able to withstand winds of up to 120 mph, which meets the building code requirements of many states. This type of roof is recommended for use in the sunbelt regions of the United States, although it can be used for small roof areas in colder climates.

Flashing

There is more to flashing than meets the eye. A professional flashing job has roofing paper or felt under the exposed flashing, whether it be some kind of metallic sheeting or roll roofing. And not all flashing is visible to the home inspector. Shingles may cover the ridge, or peak, on an asphalt-shingled roof but there will be copper, tin, or other flashing underneath.

Seasonal changes in temperature and the elements in general are rough on the roofing cement frequently used at the edges of flashing. A new coating of the cement or flashing compound may seal the edges as good as new. Small holes and cracks can be treated in the same way.

Around Plumbing Vents: All roofs having protruding plumbing vents require a piece of base flashing that is approximately 6 inches wider than the diameter of the plumbing vent itself. Since most plumbing vents through the roof (VTR's) are 5 inches in diameter, flashing units for VTR's are normally 12 inches square. This base flashing has a $\frac{1}{2}$-inch waterleg return on each side. The base flashing is cut to fit neatly around the vent pipe, and it is then sealed around the circumference of the pipe with a special sealant provided. A rubber gasket is often provided to fit tightly around the VTR. Electric service masts also require this same type of flashing.

The major parts of a roof that require flashing are shown in Figure 5-11. These include:

- Valley flashing.

- Around dormer windows.

- Around VTR's or other pipes protruding through the roof.

- Behind gutters.

- Chimneys (Figure 5-12).

Replacing flashing can be a major undertaking as shingles or other roofing material will have to be removed first to get at the edges of the flashing where the nailing or sealing has been done.

If the damaged section is in a valley of a shingled roof, it may be best to recommend patching the flashing rather than replacing the entire strip . . . unless, of course, the old flashing is hopelessly beyond repair. A square of sheet metal, of the same kind of metal as the old, is cut. This piece is then creased diagonally and folded to the angle of the existing flashing. The square should be large enough so that when folded in this way the tips on each side will fit an inch or two under

Roofs

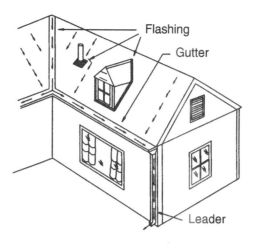

Figure 5-11: The major parts of a roof that require flashing.

the shingles at the sides. When in position, the patch will be diamond shaped, except that the tips at the sides will be under the shingles. Before the patch is positioned, it is coated with roofing cement and then pressed into place.

In proper flashing around a chimney, the edge of each piece of flashing bends into the mortar between the bricks. Each piece of

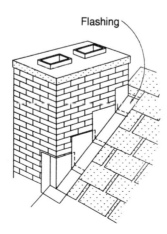

Figure 5-12: One type of chimney flashing.

flashing runs at least four inches up the side of the chimney, and each extends at least six inches under the roofing.

If you find that the house you are inspecting requires flashing replacement, a professional roofer should be called in — even for only minor repairs. Improperly installed flashing will leak, and this can eventually ruin the structural members as well as interior finishes.

A home inspector is normally not required to climb onto a roof for a close-up inspection. Rather, the experienced home inspector can tell much about a roof from the ground and inside the attic area, as will be explained in detail later on in this chapter.

Roof Structures

You should already know how to describe the various styles of residential roofs. However, there is more to a roof than meets the eye. For example, immediately under the top layer of roof covering (shingles, shakes, etc.) is an underlayment, usually consisting of asphalt impregnated felt building paper. Then comes the "deck," a substrate consisting of wide boards or sheathing. All of the above roofing layers are supported by a structure consisting of rafters and/or trusses. The best roof is only as good as these supporting members.

The area between the underside of the roof deck and the ceiling below is called the "attic," and the attic is where you can tell a lot about the condition of the roof above.

RAFTERS

In general, a rafter is a wood or metal support member on a sloping roof as shown in Figure 5-13. The basic rafters are:

- *Common rafter:* runs from ride to top plate.

- *Valley rafter:* runs at the intersection of two downward sloping roofs.

- *Hip rafter:* Runs from ridge to top plate at roof ends to form a hip.

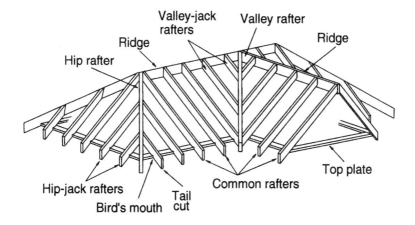

Figure 5-13: Various types of wood rafters.

- *Jack rafter:* Runs from the ridge to a valley rafter (valley-jack rafter) or from the hip rafter to top plate (hip-jack rafter).

- *Cripple rafter:* Does not touch either the ridge or top plate. Runs from hip to valley (hip-valley cripple) or valley to valley (valley cripple).

In rigid frame construction, as may be used in farm buildings or storage buildings around the home, the horizontal support part of the rigid frame is called a rafter.

Trusses

A truss is a prefabricated or manufactured roof support member that obtains its internal support by cross braces called *webs*. Trusses can be designed to fit any roof situation. Flat floor and roof trusses are also widely used. Metal connectors, called "truss clips," are used in fastening the parts of a truss together.

Attics

As mentioned previously, an attic is the space above the ceiling of the top floor and below the roof rafters. This is the area where the home inspector can tell a lot about the condition of the roof above, but it is also a place that requires adequate ventilation. In fact, proper attic ventialtion can lower attic temperatures as much as 30 degrees or more, resulting in a much cooler house during the hot summer days. In addition to lowered temperatures, an attic fan provides a constant flow of fresh air for easier breathing and greater body comfort.

If a powered attic intake fan is used, provisions must also be provided for the air exhaust.

The following table lists the approximate size of exhaust opening needed for various types of openings and sizes of exhaust fans.

Fan Size (inches)	Full Opening	Wood Louvers	Metal Louvers
22	11	16	14
24	12	18	16
30	18	27	24
36	26	39	33
42	36	54	47
48	46	69	60

INSPECTING THE ROOF

The home inspector is **not** required to climb onto the roof and walk on the roof to observe every minute detail. Doing so could cause damage to the property, not to mention the risks involved. Some inspectors disagree with this policy, but wood and clay shingles are easily damaged, and although your inspection agreement may state that you are not responsible for damages beyond the amount of your fee, anyone can sue anyone for anything and any amount. Therefore, it is recommended that you don't walk on the roofs of homes that you are inspecting. You can see much from the top of a ladder without actually climbing onto the roof; opening and looking out of a dormer window is another way to view much of the roof without walking out on it. Good quality binoculars can give you a "close" look at roof coverings from the ground.

In general, the home inspector should cover the following items when inspecting the roof of a house:

- Identify the style of roof.

- Identify the type of roof covering.

- Report the condition of the roof covering.

- Report the type and condition of roof flashing.

- Note any skylights, chimneys, VTRs or other items that protrude through the roof, such as electric service masts.

- Report the type and thickness of attic insulation.

- Report the type of attic ventilation.

- Report any signs of leaks or abnormal condensation in the attic.

- Report any tree limbs that could fall onto the roof and do damage at a later date.

- Report the procedures used to obtain the above information.

The first thing you want to do when getting to the home site is to observe the roof line and the style of roof, make the appropriate notes on your checklist. At this point, you should be able to readily recognize the roof style; that is, gable, hip, flat, shed, gambrel, etc. Then note if the roof is clean or not. Leaves and other debris on the roof tend to hold water that can promote aging of the roof covering — especially in roof valleys. Also note if any repairs are evident, and if possible, the approximate age of the repairs. Repairs indicate a problem in the past, and if repairs are readily visible, always carefully check for interior damage that may have been done due to these roof defects and have not yet been repaired.

At this point, use binoculars from ground level to observe any roof flashing around VTRs, chimneys, roof valleys, and the like. The binoculars should also give you a good indication of the condition of the roof covering. Also note any sags, bubbles, and the like that you might see.

Asphalt Shingles

A description of asphalt shingles has already been covered. Now let's see what you should look for when inspecting a home with this type of roof covering.

The majority of homes built since World War II utilize asphalt shingles as a roof covering. This type of covering will last from 15 to 25 years, depending on the quality. So, when observing an asphalt roof, try to determine the age of the shingles so you can estimate the remaining life of the existing shingles before replacement becomes necessary.

Dark places on asphalt shingles indicate that the surface granules have worn away. These granules may also be seen on downspout splashblocks and on the ground. This is a sure sign that the roof is becoming aged, and a replacement is due within five years or sooner. Also note if any shingles are broken, cracked or curled. Any of these conditions can allow water to seep underneath the top roof covering, and if so, trouble is not far away. Any of these signs are also a sure indication that either the roof is old or it was not installed properly with good quality materials.

If any shingles are missing, you can expect more to be missing shortly. The wind will begin a chain reaction that will remove other shingles next to the ones that are missing. Therefore, any missing shingles should be replaced by a professional roofer immediately.

Look at the roof flashing next. Flashing should be around any items that protrude through the roof such as vent pipes, electric service masts, chimneys, skylights, and similar items. Also look for flashing at roof valleys, roof ridges, around dormer windows, and the like. Defective flashings or the sealant around them are possible sources of leaks. You should also check to see that the flashing embedded into the mortar on brick chimneys is still intact. Loose mortar will allow moisture to get behind the flashing and seep into the interior of the home.

The inspection of asphalt roll roofing is similar to checking asphalt shingles; that is, when roll roofing shows signs of shedding its granules, it's a sure sign of aging. Once the granules are gone, sun rays attack the asphalt, causing the shingle to eventually expose the reinforcing mat. Moisture, in turn, enters the mat along the fibers leads which causes further breakdown of the shingle.

Also look for improperly installed roll roofing; that is, with overlapping done in reverse order — the lower course lapped over the upper. Such roofs will quickly leak, and must be corrected before the roof gets your stamp of approval.

Built-Up Roofs

When inspecting homes with flat roofs, a low spot where water collects is potential trouble. Look for bubbles in a tar-and-gravel roof. This is a sure indication that moisture has worked its way under the roofing felt.

Ultra-violet radiation breaks down the surface, while thermal expansion causes cracks. Both of these actions will eventually break down the surface material to expose the fibers underneath. Once these fibers are exposed, moisture will enter under and between roofing layers by capillary action. This moisture, when heated by the sun, causes expansion which is indicated by air pockets or bubbles in the roofing materials.

Wood Shingles and Shakes

Western red cedar or redwood shingles should last for 20 to 25 years when used on roofs with a 4 to 12 pitch or greater. The steeper the pitch, the longer the roof will last. Some vacation cabins and outbuildings utilizing cedar shakes have maintained a leak-free roof for over 50 years, but this is the exception rather than the rule.

When inspecting roofs with wood-shingle covering, look for missing shingles, cracks or splits in shingles — especially those that lie directly over or under gaps in the adjacent courses. Any of these conditions can cause the roof to leak.

Look for signs of moss or algae on the roof. This is an indication that these areas are damp. Try to determine why. Detecting insects and rot will be covered in a later chapter.

Metal Roofs

Metal roofs were common for use on houses during the first half of this century — especially those homes utilizing wood heat. Although noisy during a rain storm (unless heavily insulated), metal roofs have proven their worth for fire protection against chimney sparks. They are also very durable when properly maintained with a protective coating of paint every few years. Many such roofs are over 75 years old and still show no signs of leaking or deterioration.

When inspecting metal roofs, look for rust spots that could eat through the metal. Also be aware that metal roofs attract moisture during cold seasons, and during these times, the inside of the roof may be continually damp. Consequently, suitable ventilation must be provided to keep the wood framing from rotting.

Also look for signs of repairs, indicating trouble in the past. Repairs may be in the form of soldered patches, asphalt shingle patches, or a heavy coat of roof sealant. Once you get into the attic you will be able to tell more about the roof by looking for water stains and other tell-tale signs.

Tile Roofs

Tile roofs have been known to last for hundreds of years and require very little maintenance. The home inspector's first concern when he encounters such a roof is to make sure the structure underneath is adequate. Tile and slate roofs are heavy, and both require more support than most other types of roofs.

Sometimes clay tiles will become broken when walked on, or knocked off completely. Such damage is usually done when some other type of installation or maintenance is done to the home; that is, tv antenna installation, chimney cleaning, or installing plumbing vent pipes. If any tiles are missing, they should be replaced immediately because a vacant spot will allow high winds to disarrange other tiles in the area.

A Closer Look

After a careful check of the roof covering and flashing, you should look at the eaves and soffits for obvious signs of rot or decay — especially behind gutters. Gutters are prime trouble spots in cold climates because freezing rain can lift shingles and allow moisture to seep underneath. Push the tip of an ice pick or screwdriver into the wood on the edge of the eaves to check for wood rot.

Gutters are installed on eaves to collect rain or melting snow from the roof and channel it into downspouts that direct the water away from the house foundation. All gutters should be examined for leaks, cracks and weak spots. They should be clean, without sag, and firmly secured. Leaking gutters can cause damage to siding and trim and allow water to enter the basement or else weaken the foundation.

Wood gutters were once the standard, but are seldom encountered on homes built in the past 50 years. Galvanized metal gutters are probably the most common, but must be kept painted — inside and out — to prevent deterioration. In the past couple of decades, aluminum and vinyl gutters and downspouts have entered the picture, and these types are practically maintenance free, except for a cleaning once or twice a year.

Unless obvious rust spots are visible on either the gutters or downspouts, the best time to check these items is during a heavy rain. Look for water flowing over the top of the gutter, leaks at seams and other places. Also look for sagging gutters, damaged downspouts, corrosion and evidence of loose fasteners. A splashblock should also be provided at each downspout.

When the roof drainage system is not functioning properly, the house structure will eventually become damaged, in the form of leaking and flooding — the very things gutters and downspouts are designed to prevent.

The checklist in Figure 5-14 will help you in your inspection of the home's roof.

THE ATTIC

Much information about the home's roof can be determined from an attic inspection. If the attic area is unfinished (and most are), use a flashlight to carefully examine the underside of the roof deck and around each rafter. Water stain is a sure sign that the roof is leaking or once leaked. Wood that is covered with mildew is an indication of a severe leak. Look for discolored wood members and then try to determine the source of the problem. Also use your ice pick or

INSPECTION CHECKLISTS — ROOFS							
Circle the appropriate answers below:							
Type of roof:	Gable	Gambrel	Hip	Flat	Shed	Other	
Type of roof covering:	Tile	Metal	Fiber-glass	Roll Roofing	Other		
Type of roof flashing:	Tin	Copper	Other				

		Yes	No
1.	Are there any dormer windows?		
2.	Are there any loose or missing shingles?		
3.	Are there any smooth spots on the roofing material?		
4.	Is every opening in the roof properly flashed?		
5.	Are there any signs of leaks on the underside of the roof?		
6.	Are any boards on the gable ends loose or in need of repair?		
7.	Are any boards on dormer windows loose or in need of repair?		
8.	Are there any signs of water seepage around the chimney or other roof openings?		
9.	Does the roof sag?		
10.	Are the soffits in good condition?		

Overall condition of the roof and related structure:	Good	Fair	Poor
Defects:			
Repairs needed:			
Quality of construction:			

Figure 5-14: Home inspection checklist for residential roofs.

screwdriver to probe for suspicious-looking spots for possible wood rot.

While in the attic, note the type and sizes of attic ventilators, grilles, etc. Also note the type and thickness of attic insulation. Details on attic ventilation will be covered in Chapter 11 — *Heating, Ventilating, and Air Conditioning*.

Chapter 6
Chimneys and Flues

The first log homes built in America usually had one large fireplace that was used for heating, cooking, and lead bullet casting. As construction techniques became more sophisticated, colonists wanted larger homes with a fireplace in each room. Some designs used a chimney at each end of the house to accomplish this, while others had a fireplace in each room, all feeding into one central chimney.

In the middle of the 18th century, Benjamin Franklin developed a cast-iron, wood-burning stove with metal surfaces that radiated heat in all directions, not in just a single direction the way a fireplace does. The stove sat on heavy legs and could be placed anywhere in a room as long as its smoke pipe could reach the chimney. The average fireplace has an efficiency rating of approximately 10 to 15%, while the Franklin stove has a rating of from 25 to 30%. This meant more heat for less fuel. Thus, free-standing iron stoves became a big hit and were used in nearly every household in America well into the twentieth century. Consequently, homes built before the 1930s usually had several chimneys. Even after coal-, gas- and oil-fired furnaces came into play during the first part of the twentieth century, each home still had to have at least one chimney for these heating devices to operate. Most of the homes also had a fireplace in the living room and two flues were usually installed in one chimney structure — one for the fireplace and the other for the furnace.

With increasing heating fuel costs, homeowners are turning more and more to wood for supplemental heat. In doing so, they are reducing their fuel bills substantially, especially if these homeowners are fortunate enough to have a woodlot or cheap source of fuel.

Chimneys in most older homes are probably in need of some repair. This chapter is designed to show you how to identify the various types of fireplaces and chimneys, how to evaluate the condition of each, and make recommendations on chimney repairs and maintenance.

CHIMNEY INSPECTION

Chimney flues for wood stoves should be ample in size and carried as nearly straight as possible from a point near the stove to above the highest projection of the roof. They should preferably be independent, having no connection with other flues (furnace, fireplaces, etc.) or openings, and always of the same area from top to bottom. A well-jointed tile flue with tightly cemented joints is a must — not only for proper operation, but for safety as well.

When looking over an existing chimney with a woodburning stove connected to it, the following observations should be made:

- See that no openings exist into the flue, either above or below the stove's smoke pipe.

- It is best for the division walls of a chimney, if it contains more than one flue, to be carried up to the top of the chimney, and down to the bottom of the chimney, so that each flue is independent of the other throughout the entire length.

- The area of the chimney flue should be maintained full size throughout its entire length and be free from all obstructions such as loose brick and mortar that might become lodged in it. An offset in the flue should have increased area to overcome the added friction of the offset.

- Look for oily substance like *creosote* on the inside of the flue. Creosote is usually caused by burning green wood in a tightly closed stove. This substance can ignite and is

Chimneys and Flues

often the cause of chimney fires. A mirror may sometimes be used to detect obstructions and the general condition of the flue.

- The chimney should extend above the highest point of the roof or other immediate surroundings such as adjoining buildings, hills, or trees. If not, a poor draft can be expected.

- The smoke pipe should not project too far into the chimney.

To test for other openings or leaks in a chimney, twist a newspaper and light the end. Then hold the lighted newspaper against the opening where the stove's smoke pipe will enter. If part of the flame goes downward, there is a leakage below the smoke pipe entrance (Figure 6-1).

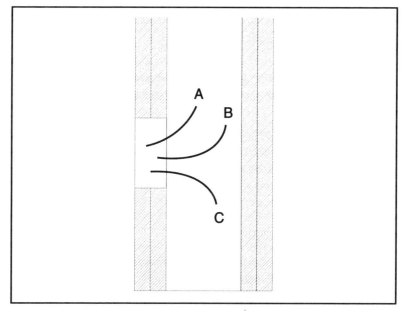

Figure 6-1: The existence of another unwanted opening at the bottom of a flue, or a leak of any other type, may be readily detected by twisting a newspaper, and lighting and holding it against the opening where the smoke pipe enters the chimney flue. If part of the flame goes downward (C), this indicates a leak below the thimble opening.

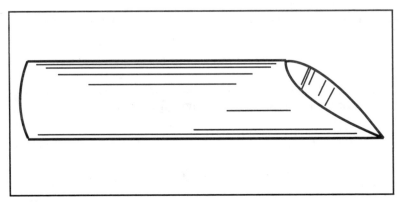

Figure 6-2: Method of cutting stovepipe for better results when the flue extends below where the pipe enters the chimney.

In most cases, when the test indicates a leakage below the opening, it is caused by two flues being joined together at the bottom for cleanout purposes — using only one cleanout door for two or more flues. New cleanout doors should be provided at the bottom of each flue. The existing or middle cleanout door is used to seal off the two flues with cement.

Where it is necessary to have the flue extend below where the stove's smokestack or pipe enters, unsatisfactory results may be avoided by giving the end of the smoke pipe the form indicated in Figure 6-2. Care, however, must be taken to see that the pipe does not become turned the wrong way. If you run into a situation where the chimney seems to not draw properly, check the smoke pipe end before making your final report.

Wood Stove Inspection

While you're checking the chimney, it wouldn't hurt to give any wood stoves a check. Wood stoves listed by Underwriters' Laboratories (UL) or approved according to UL standards come from the manufacturer with instructions on ideal use and maintenance. Most of the approved woodburning stoves are airtight and can be closed off in the event of a chimney fire. While most lending institutions will not be too concerned with the type of wood stove — used for supplemental heat in a home — insurance companies will. Therefore, as a home inspector, you may be asked for detailed information about any wood-burning stoves in a particular home.

Creosote, the main ingredient in flue buildup, comes from the burning of soft or green wood. Pine is the most notorious and should be avoided as much as possible except for kindling. Another mistake many people make in using wood stoves is purchasing a wood stove larger than needed to heat a particular area. For example, a stove designed to heat a 2,000-square-foot area operating in a 1,500-square-foot area will burn at a lower rate than required. There is more chance of creosote buildup.

The Virginia Uniform Statewide Building Code recommends the following suggestions when installing woodburning stoves. The home inspector should be aware of this and other codes when inspecting this part of any home.

- Be sure there is an 18-inch clearance between the stove and floor with a 24-gauge steel pad placed beneath the appliance.

- Check the stovepipe regularly for corrosion and holes. Replace if necessary.

- Install the small end of the stovepipe down, or pointed toward the stove.

- Avoid long horizontal runs, and be sure the pipe is securely supported.

- Minimize the number of elbows.

Figure 6-3 on the next page shows several correct methods of installing woodburning stoves.

Chimney Inspection

If the chimney is for a conventional fireplace (with no metal stovepipe connection), start with a clean fireplace; that is, all ashes should be removed. All loose soot from the walls of the fireplace, ceiling, and around the damper door should be removed.

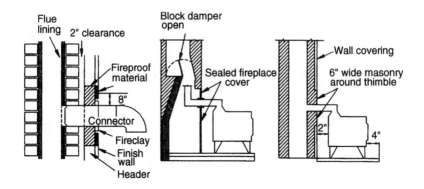

Figure 6-3: Correct methods of installing woodburning stoves in several situations.

For a masonry chimney to which a stovepipe is attached, a vacuum cleaner may be used to remove all soot from the outside of the thimble opening to where it attaches to the chimney flue. A wire brush may be necessary to remove some of the soot and other debris. Again, make sure your wall is protected directly under the thimble opening. Newspaper attached to the wall with masking type will normally do the trick.

Flue: A mirror and a flashlight will allow you to examine the chimney flue. One of the small mirrors like those in ladies' compacts is ideal. Attach this to a small slat to project it back into the chimney through the thimble opening, and then use a flashlight as shown in Figure 6-4. You should have a full view of the chimney from the thimble opening to the top. If enough light comes in through the top of the chimney, the flashlight may be necessary.

You should have a pencil and pad handy to take notes as you view every inch of the chimney flue. Some inspectors

Figure 6-4: Using a mirror and flashlight to examine inside of chimney.

use a small cassette tape recorder. When they want to note points of interest, they merely talk into the tape recorder, which is continuously running during the inspection, and any message is recorded for later reference.

This mirror/flashlight inspection will allow you to detect animal and bird nests that must be removed before the chimney can be operated safely and effectively. Note the condition of the flue lining and the amount of soot and creosote present, the condition of the flue joints, etc. This inspection will also let you know if the chimney is lined with flue liner or merely built out of bricks. If the latter situation is the case, you will want to recommend the installation of flue lining if the chimney is to be used.

Joints: The outside of the chimney comes next. Usually a quick visual inspection from the ground will reveal the condition of the overall chimney. If the masonry joints look tight, chances are all of them will be tight. If some of the joints at ground level to eye level are deteriorating, you'll probably have the same thing happening farther up. Any loose bricks should be noted in your report. If some of the joints seem in poor shape, you should recommend that the entire brick joint be scraped out. New mortar should then be applied to insure a tight fit.

Should several of the joints show deterioration, you should use a ladder and inspect all joints carefully. The best time to perform this inspection, if possible, is to wait until dark and have a helper shine a relatively powerful light up or down the chimney flue; then carefully inspect each joint to make certain you can't see any light through the joints. If so, then the joint should be repaired. You might want to carry a piece of chalk with you as you make the inspection to mark the various defective areas. Where the chimney runs through an attic or other spaces within the house, you should inspect these places the same way.

Leaks: Another consideration during the inspection is the distance chimneys and chimney openings (thimbles) are from wooden structural members. A roaring chimney fire can reach temperatures in excess of 3000°F. Framing members within 12 inches or less of an opening (thimble, defective joint, etc.) can ignite, causing an entire home to burn. One place to be especially cautious is where a chimney thimble opening is located near a combustible ceiling. A chimney fire can melt the metal stove pipe or cause it to be hot enough to ignite a combustible ceiling that's too close to the opening. Such conditions can be corrected by placing a sheet of noncombustible material directly above the chimney opening.

The thimble opening should provide a relatively tight fit for a stovepipe — tight enough so that drafts will not cause smoke to flow out the sides and fill a room. Furthermore, the pipe should be secured tightly enough so it does not come loose due to vibrations or an accidental jarring of the stove. Do not, however, have the opening too small as the pipe will crimp when it is fitted into the opening. This will obstruct the flow of smoke and hot gases and may result in creosote buildup.

Many masonry chimneys are built flush with the inside wall and then plastered over the brickwork to match the adjoining walls. After a house settles, there will be an obvious crack between the chimney and the adjoining walls. This crack should be sealed.

You'll find that chimneys (especially those that are built on the interior walls of a house) are trouble spots for leaking around the chimney flashing. This leaking will rot wooden members and spoil ceiling plaster and the wall plaster in the immediate area.

Footing: Most chimneys and fireplaces have footings at least 2 feet deep. Consequently, chimney footings usually offered no serious structural problems, but it's still a good idea to check the footing when inspecting the chimney. If repairs are necessary, professional contractors should do the job.

Chimney Cap: The chimney cap is usually made of poured concrete to keep moisture, dust, and other foreign matter out of the masonry. Chimney caps will develop cracks that allow moisture to seep into the seams of the masonry work. The joints may be weakened. Inspect the cap and suggest necessary repairs.

Chimney Hood: Consider chimney accessories such as a hood to prevent gusts of wind or drafts from blowing down the chimney. When considering such a device, however, make certain that the cross sectional area of the flue is maintained. The total area left open must be at least equal to the cross sectional area of the flue. Some builders prefer just a trifle more. Some choose to close the hood on two sides and open it on the sides to the prevailing winds. This promotes a sustained draft while eliminating downdrafts.

Consider the effect of two flues in the same chimney during your inspection. One should be higher than the other to prevent the drafts

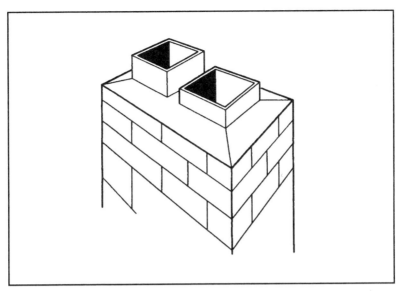

Figure 6-5: When two flues are in the same chimney, one should be higher than the other to prevent the draft from one flowing down the other.

from one flowing down the other (Figure 6-5). Draft short-circuiting may also be corrected by installing a hood and then separating the two flues with a wythe as shown in Figure 6-6.

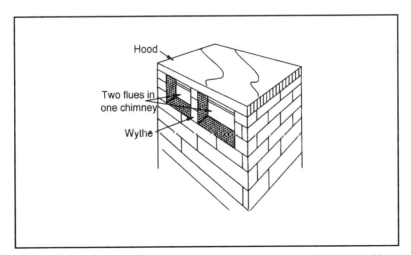

Figure 6-6: A hood with a wythe installed is one way of dealing with chimneys that have two flues in them.

FIREPLACE INSPECTION

If you are inspecting a fireplace with a woodburning unit installed within the fireplace, or a woodburning stove is connected to the fireplace flue with the fireplace opening sealed off, the stove should be removed before the inspection begins.

You should begin your inspection by checking the fireplace footing. Since the chimney is the heaviest part of most homes, it must be built on a solid foundation of concrete. The minimum size should be 12 inches thick and project 6 inches beyond the chimney. If the foundation is not of adequate size or in need of repair, only a professional should do the work.

One part of the fireplace that is often overlooked is the ashpit. Although not absolutely necessary, it is a great convenience for removing ashes without getting them all over the living room floor. The greatest faults include a rough passageway to the cleanout door, which promotes ashes clogging on their way downward. The worst problem is allowing water to leak into this space. When water is mixed with ashes, a cement-like mass is formed that makes the ashes extremely difficult, if not impossible, to remove.

When inspecting the ashpit, make sure the walls and floor (just below the cleanout door) are smooth. Then make sure the area is watertight. This means no leaks in the masonry and a gasket on the cleanout door, especially if it's located on the outside of the home and is exposed to the elements. Make sure the back of the fireplace is sloped to reflect heat outward and to direct smoke and hot air toward the throat of the fireplace.

See that the damper is functioning properly. The size of the opening is controlled by raising the valve plate with a lever or screw. A damper door that will not close completely will suck enormous amounts of hot air up the chimney when some other type of heat is being used.

The damper is another part of the fireplace that is often installed incorrectly. The lintel across the front of the fireplace should be separate from the damper frame. Sometimes the damper frame is used as a lintel (Figure 6-7). If this is the case, the easiest solution is to lower the front opening of the fireplace by adding a hood (Figures 6-8 and 6-9).

Chimneys and Flues

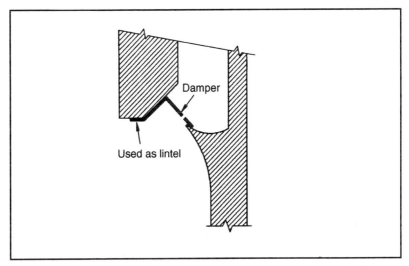

Figure 6-7: Some dampers are used as lintels, which can result in smoking fireplaces.

Another fault is that the damper is sometimes shorter than the width of the fireplace opening; it should be the same width. You can correct this error in two ways; replace the damper with a longer one or make the firebox narrower by building up the inside with firebricks.

The fireplace should have a smoke shelf to keep cold air current from flowing down the chimney. You can help a defective smoke shelf

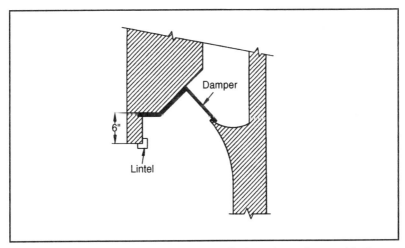

Figure 6-8: The damper should be installed as shown here, with the lintel about 6" below the damper door.

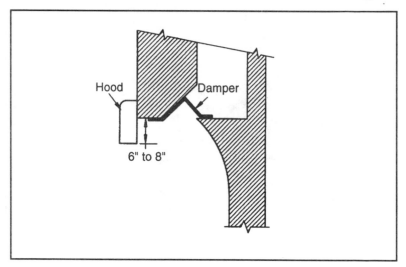

Figure 6-9: One way to correct a faulty damper door is to install a hood across the firebox opening to make the lintel 6 - 8 inches.

somewhat by installing a hood at the top of the fireplace opening and also at the top of the chimney.

Chimney Fires: Although most chimneys will probably withstand the heat of a chimney fire, flames may pass through a crack into the walls, or burning flakes of soot may ignite the roof. If a chimney breaks into flame, the fire department should be notified at once. While waiting for the firemen, or if the fireplace is beyond fire service, throwing salt or baking soda on the fire will help. Dousing the roof with a garden hose will also help prevent the roof from igniting.

An inspection checklist for chimneys and flues appears in Figure 6-10.

Chimneys and Flues

INSPECTION CHECKLIST — CHIMNEYS AND FLUES				
Total number of chimneys:					
Total number of flues per chimney					
Total number of fireplaces					
Total number of other heating devices requiring a flue					
			Yes	No
1.	Are all chimneys flue-lined?		_____	_____
2.	Are any bricks missing from the chimney structure (both inside and outside)?		_____	_____
3.	Do all fireplaces contain dampers?		_____	_____
4.	Are all space-heating devices UL rated?		_____	_____
5.	If more than one flue is installed in a single chimney structure, is one flue higher than the other?		_____	_____
6.	If more than one flue is installed in a single chimney structure, are the flues divided by a wythe?		_____	_____
7.	Are spark arresters used on all flues with woodburning equipment attached?		_____	_____
8.	Are there any signs of smoking fireplaces; that is, blackened masonry around fireplace opening, etc.?		_____	_____
9.	Are all fireplaces provided with sufficient hearth area in front of openings?		_____	_____
10.	Are all woodburning stoves located at least 18 inches from any combustible materials?		_____	_____
		Good	Fair	Poor
Overall condition of chimneys and flues:					
Overall condition of fireplaces:					
Overall condition of other space-heating devices:					

Figure 6-10: Inspection checklist for chimneys and flues.

Chapter 7
Interior Finishes

Plastering whole walls in any home is becoming a thing of the past. Drywall or gypsum board with taped joints has all but replaced plastering in modern building construction. However, the majority of homes built prior to World War II were plastered. Furthermore, some quality homes still use the plastering methods. An expert drywall job is hard to tell from one that is plastered, but the experienced building inspector can tell the difference in drywall and plastered walls with a tap of the hand, even under wallpaper. The drywall tap will have a hollow dull sound, while a plastered wall will have a sharp, higher-pitched sound. Obviously, it takes some experience to detect these different sounds, but once you gain this knowledge, it's like riding a bicycle: once you learn, you'll never forget.

Which of the following reports do you think would sound better to the people hiring your services:

- "This house has wallpapered walls."

or

- "This house has good wallpaper over sound plastered walls. The one exception is a new wing built on the rear of the house which has been finished in painted drywall with taped joints. Workmanship is of the highest quality."

Don't you think you would have an easier time collecting your fee, plus a greater chance of more business, if you submitted the second report rather than the first?

Since the condition and type of wall finishes will affect the value of any house, you must know what to look for and then how to describe what you found. This chapter is designed to do just that. Upon completion of this chapter, you should be able to accurately describe most types of finished walls in residential construction.

INSPECTING THE INTERIOR

When inspecting a home's interior, you will observe such items as walls, ceilings, floors, steps, stairways, balconies, and railings. You will also inspect cabinets and counters in the kitchen and bath. Do not overlook windows and doors, as well as the lighting fixtures around the home.

As you enter the home, inspect the entry door and examine the hardware (hinges and locks). Check the wall light switches and also note if the foyer's lighting fixture is in good condition.

Continue to walk through the home, observing the walls, ceilings, and floors of each room. Examine and operate all doors and windows, including closet doors. As you enter each room, operate the light switch to make sure it functions properly. Note the condition of any permanently mounted lighting fixtures. See Figure 7-1. Table lamps and other portable lights are not the home inspector's concern.

Figure 7-1: Always check the condition of each lighting fixture in the home. Note any broken glass or inoperative lamps.

Interior Finishes

Identify the floor, ceiling, and wall finishes. Is the floor carpeted, hardwood, or some other finish? How about the walls — are they plastered, drywalled, or paneled? In your report, note any walls that are covered by wallpaper.

Look for cracks in the ceiling and walls; look for stains, water damage, powdery areas, broken plaster, walls that are not plumb, and sags in the ceiling. Popped nails in a plastered or drywall wall or ceiling could indicate a structural defect.

When checking the floors, look for areas that are not level, damaged flooring, bumps in the floor, and warped or loose floorboards. Identify any floor covering, such as linoleum, tile, slate, etc. You can carry a tennis ball with you to roll across the floors to see if they are level. A conventional carpenter's level can also be used.

You should identify the door types; that is, panel, flush, louver, sliding, etc. Also identify the material from which the doors are made. Then check their condition for:

- ease of operation
- sticking doors
- noisy doors
- cracks or broken-out corners
- condition of hardware on each door
- scratches
- squareness of door frames

Treads on stairs should be solidly fastened (see Figure 7-2 on the next page); risers should be the same and unbroken. Stairs must be well-lit so there is no danger of tripping on them. Don't forget understair closets during your inspection.

Identify window types, including the type of glass used. If insulating glass is used, make a note of this. Storm windows on the outside of the home usually indicate that conventional glass is used rather than insulating glass. Don't forget to check the caulking on the outside of

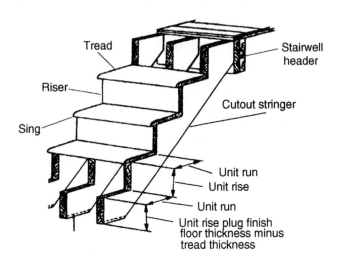

Figure 7-2: Treads on all stairs should be securely fastened.

each window for peeling and cracking. Improper caulking will allow moisture into the home, not to mention the heat loss and heat gain through air infiltration.

INSPECTING TRIM AND BUILT-INS

Trim has a number of meanings in the building construction industry, from wrought-iron hinges to cabinet veneer. In all cases, however, the term *trim* is used to indicate something that is used to finish off something else.

Molding is one type of trim. It will be found in virtually every house you will inspect — around windows and doors, and where the floor meets the walls. Some homes will have additional kinds of molding: cove molding, where the walls meet the ceiling; inside and outside corner molding; and chair rail or picture molding along the walls or atop wainscot panels (Figure 7-3), just to name a few.

Most naturally finished interior trim is made of either Ponderosa pine or white pine. However, you will also find molding designed for painting milled from yellow pine, gum, poplar, and similar woods. Trim for staining is also available in oak, walnut, cherry, birch, and

Interior Finishes 131

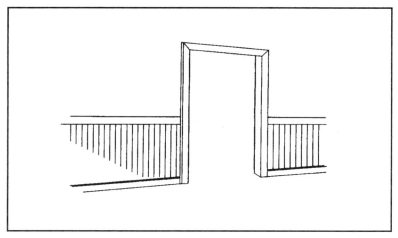

Figure 7-3: Wainscoting is frequently used for room walls; check the molding where the panel end and the plain wall starts.

mahogany. All should be free of knots and other defects. Although wood is the most popular type of molding, you will also find vinyl and hardboard moldings, especially in recently built, medium-quality homes.

First Things First

When you enter a home that you are inspecting, first determine if the molding is symmetrical or nonsymmetrical. Some ceiling molding, for example, has one surface wider than the other. The part that fits up against the ceiling may be 1½" wide, while the part that goes next to the wall is 2" wide. Such molding gives a decorative effect that is usually quite striking, but is difficult to fit in place. If the installers were professionals, the corner joints should be perfect; if not, you will see uneven joints, and perhaps spots where the installers' mistakes were covered up with some type of wood filler. The reason: most molding is mitered to a 45° angle (like the corners of a picture frame) and the cutting gets more difficult if unequal surfaces have to be matched up. Such molding, and the way it was installed, can tell the home inspector a lot about the overall handiwork of the entire home.

Spot-check all molding to see that the miter joints have been carefully made, since nothing detracts from the finished appearance of a trim job as much as poorly fitted trim.

There's a slight trick to fastening molding in place. Trim does more than just dress up the area. It's intended to hide the joint (which can sometimes be sloppy) where two surfaces meet. Molding also serves to hide the gap should the materials contract and leave a wider opening. For this reason, the molding should never be fastened to the ceiling, the floor, or the walls. Instead, the finishing nails should be driven through the mold at a 45° angle so they will go past the surfacing and into the studs beyond. By doing so, the walls, floor, and ceiling are free to shift slightly without dragging the molding along.

Determine if any nails in the molding have been countersunk below the surfaces and the holes plugged with wood filler. During installation, once the wood filler has been applied, the molding should be sanded smooth and finished in a color that matches or contrasts with the decor of the room. If any strange conditions are found, you should mention this in your inspection report.

Check to see that shelves in cabinets and bookcases are secure and have no cracks. If the shelves are of the adjustable type, check to see that the adjusting hardware is in good order. Many built-in bookcases are veneered with some type of wood grain — walnut, cherry, mahogany, etc. Since veneer is glued onto the wood, sometimes this glue will become dry and hard and will no longer hold. Check all corners of veneered wood to see that the veneer is not peeling. If peeling is found, your report should indicate the room and object that needs its veneer reglued.

CEILINGS

By definition, a ceiling is the overhead finish of a room used to conceal the floor or roof above. A ceiling floor is the framework, including joists and covering, that supports a ceiling. Ceiling joists are the intermediate, horizontal, structural members used to support the finished ceiling material. If there is a floor above the ceiling, then ceiling joists also serve as floor joists to the floor of the room above. See Figure 7-4.

This section will introduce you to the various types of ceilings used in residential construction, how they are constructed, the common faults that occur, and how to detect them. Furthermore, this section

Interior Finishes

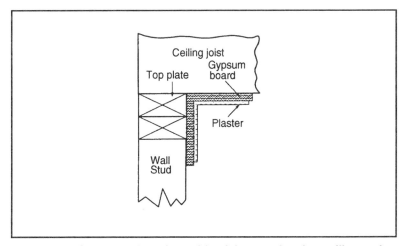

Figure 7-4: Cross-section of a residential room showing ceiling and related structural members.

describes the various ceiling finishes so that you can accurately describe them in your home inspection report.

One of the most common types of ceiling finishes in older homes (those built before World War II), is the use of plaster on wood lath. This type of ceiling was used in all living areas and sometimes in the basement and attic areas. Thin wood strips nailed directly across the face of wall studs and ceiling joists constituted the lath on which the plaster was spread. Horsehair plaster was a common type of plaster. Real horsehair was added to the base coat to keep the plaster from oozing through the wood lath. This hair can be easily detected in old plaster that has been removed from the lath.

Plastered ceilings were replaced with the advent of gypsum board. Today, plastering is time-consuming and laborious, and is therefore expensive and rarely used for new home construction. Plaster is made up of gypsum and lime or Portland cement, sand (or vermiculite or perlite), and water. The finish coat consists of gypsum gaging plaster or Keen's cement. Gaging plaster is applied before the base coat is fully dry. Keen's cement finish is applied after the base coat is dry to give a harder finish and because it is more resistant to moisture.

Today, plaster is applied over wire mesh and is commonly a three-coat application. First, the *scratch coat is* applied. It gets its name from the fact that it is scratched with a metal tool to aid in the bonding of the second brown coat.

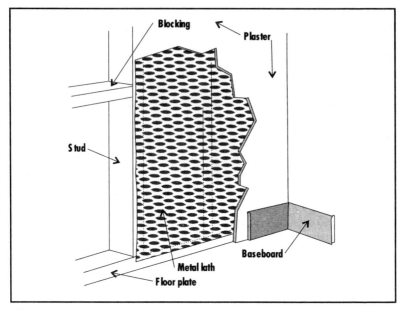

Figure 7-5: Principles of plastering.

The final finish coat dries to a smooth, white finish that can be either painted, papered, or left white. The principles of plastering over a metal lath are shown in Figure 7-5.

Gypsum Board

Gypsum board, commonly called *drywall* or *sheetrock is* the most common type of ceiling found in new homes today. It is composed of a hard gypsum center that is given a structural rigidity by surface paper bonded to both sides. It is placed directly against wall studs and nailed or screwed in place. It can also be glued directly to concrete block or nailed to furring strips.

Nails are slightly recessed into the surface of the gypsum board so that the gypsum board compresses and leaves a hammerhead indentation. Then the joint compound can be applied flush with the surface, completely hiding the nail head. Another popular method of installing the gypsum panels is with Phillips-head, gypsum-board screws. The screws are placed in a driving attachment chucked in a small, electric drill.

Wood Ceilings

Exposed: Exposed ceilings (Figure 7-6) are those that have no actual ceiling finish or the finish is limited. The underside of the floor or roof above is left exposed. This may well have been the first type of residential ceiling. Exposed ceiling structures in post-and-beam constructed residences are considered very attractive and highly desirable. When such a ceiling is used, however, much attention must be given to the color and finish of the exposed area. Builders of log houses and A-frame structures often have the exposed wood hand-sanded and then stained and varnished. Older homes, restored houses, and modern colonial or country design homes often utilize exposed beams. Such areas may have only the beams exposed, while the floor or roof boards above are covered with gypsum board or plaster and then painted. In some structures, however, both the structural beams and the flooring above is left exposed for an all-wood effect.

Figure 7-6: Exposed beam ceiling.

Exposed beam ceilings are especially attractive in older homes built of logs or post-and-beams. The roof beams, or rafters, of a sloped roof ceiling are often left exposed while the underside of the roof sheathing between the rafters is covered with insulation board and then finished with painted gypsum board.

Today, fake exposed beams are often used effectively in modern homes. The home inspector will usually encounter such fake beams in recreation rooms and kitchens.

Ceilings with an all-wood finish, once used extensively for ceiling coverings in every room, are rare today, except in larger homes where a variety of finishes is desired. Sitting rooms, dens, and screened porches are common places for decorative wood ceilings. You will find tongue-and-groove cedar boards or rough-sawn decorative wood as ceiling finishes. Many older homes with exterior porches have painted wood ceiling coverings consisting of thin, grooved, narrow wood strips with tongue-and-groove edges.

Cathedral Ceilings

Cathedral-type ceilings are sloped ceilings that generally follow the underside of sloped roof rafters. Sometimes, in order to achieve a cathedral effect an artificial framework is installed to give the appropriate slope. *Cathedral* generally implies a ceiling that is symmetrically sloped up to the room mid-point from two sides. This technique is used effectively to make rooms seem larger and less cramped. See Figure 7-7.

The term sloped ceiling is sometimes used interchangeably with the term cathedral. Actually, sloped ceiling refers to a shed slope; that is, a roof that slopes in one direction only. See Figure 7-8.

Ceiling Tile

Fire-retardant tiles are readily available and in many cases are the standard tiles for finished basements. Older tiles made of mineral and wood fibers can be highly flammable and the home inspector should make note of any such tile found.

Ceiling tiles normally come from the manufacturer in a prefinished state, although they may also be painted. Ceiling tile tends to be a good sound absorber and so is often termed acoustical tile. However, paint-

Figure 7-7: Typical cathedral ceiling used in a residence.

ing may clog the open tissues and reduce the overall sound absorbing qualities.

Fiberglass tiles have a plastic surface that is not intended to be painted. Most of the prefinishes are plain ceiling tile or bone-white, either flat or with a sheen finish.

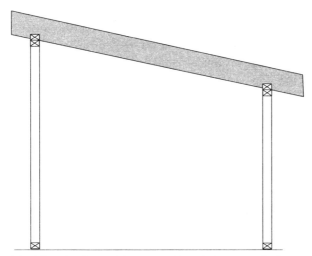

Figure 7-8: All cathedral ceilings are sloped, but not all sloped ceilings are cathedral type. The sloped ceiling shown here is more accurately described as a sloped ceiling under a shed roof.

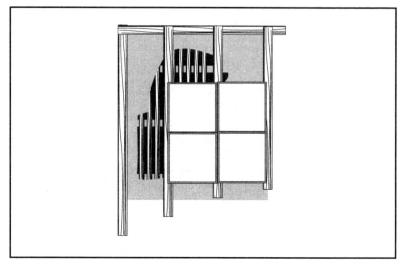

Figure 7-9: Acoustical tile installed directly over furring strips is a good covering for ceilings with broken plaster.

Most ceiling tiles measure 12" x 12" and many are designed to absorb as much as 45% of the noise that strikes them. Several installation methods have been used. On absolutely flat ceilings, tiles may be installed directly to the plaster or gypsum board with mastic cement. Ceilings that are not level, or when the tile is installed directly to ceiling joists, usually have a grid of furring strips nailed to the structure above. These furring strips can be shimmed during the installation to level the strips. The ceiling tile can be stapled directly to the furring strips with good results.

Another method utilizes a support track screwed directly to the underside of the existing structure. The finished tiles then clip to this grid without stapling or cementing.

Ceiling tiles installed on furring strips is an inexpensive way to cover a ceiling with broken plaster, as shown in Figure 7-9. When used in basements, the ceiling may be floated as shown in Figure 7-10 to conceal ductwork, plumbing pipes, and electrical wiring.

Suspended Ceilings

Suspended ceilings are ceiling-grid systems that are suspended and supported by wires hooked onto the floor or roof structure above. The

Interior Finishes

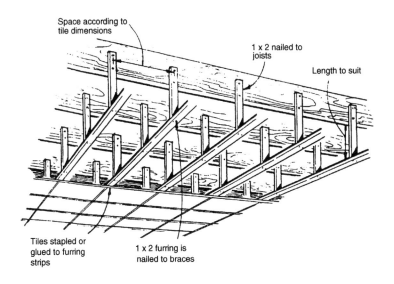

Figure 7-10: Method of "floating" a new ceiling beneath the floor joists to conceal or clear obstructions.

grid is an interlocking aluminum strip that has an inverted T-shape. Lay-in ceiling tiles are placed on the metal grid after the grid is installed. The standard grid dimension is 2' x 4', but 2' x 2' is common in residential applications. See Figure 7-11.

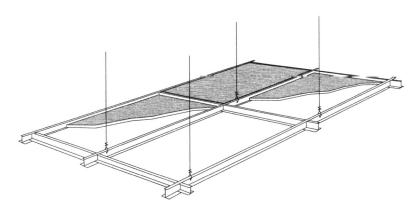

Figure 7-11: Basic components of a suspended ceiling.

The bottom of the metal grid (about ¾" wide) is usually visible, although grid systems that have a hidden grid are also common. When exposed grids are installed, a variety of colors is available to match the room.

The lay-in ceiling tiles are either 2' x 4' or 2' x 2' and fit the standard grid without any modifications except at locations where the grid is altered to match the room shape. The most common material used for lay-in ceiling tiles is wood fiber, mineral fiber, and fiberglass. Fire-retardant tiles are available and are the type that should be used in all new construction.

For residential use, suspended ceiling-grid systems are usually confined to the basement area where aesthetics are not the primary factor. If there is enough headroom (some grids can be installed with only 2" of overhead installation clearance), ductwork, plumbing pipes, electrical wiring, and similar items can be conveniently covered with a suspended ceiling. Attics are another place in the home where such a ceiling system might be warranted, provided the ceiling height is sufficient.

Luminous ceilings are another form of suspended ceiling and are sometimes used in modern residential kitchens to give a continuous skylight effect. *Luminous ceilings* utilize a supporting metal grid in which lay-in fluorescent lighting fixtures are installed. When these fixtures are energized, the entire ceiling lights up with diffused light, with no shadows.

Miscellaneous Ceilings

Trayed Ceilings: This type of ceiling is used in more elaborate homes. Dining rooms are a common space to install a trayed ceiling. It is made by placing diagonal struts at the top of a wall at the ceiling edge and around the entire perimeter of the ceiling. The ceiling finish (commonly gypsum board) is extended down over the struts. The diagonal portion of the ceiling or wall is commonly 2' x 3' wide. For this to be effective, the ceiling should be higher than normal and the room should be spacious. See Figure 7-12.

Coffered Ceilings: A coffered ceiling is used primarily in large elaborate homes with high ceilings. These are ceilings with recessed panels or the effect of recessed panels. This is accomplished by running decorative beams at right angles to each other forming a ceiling grid.

Figure 7-12: Trayed ceiling.

This type of ceiling is seldom used in modern homes due to the great expense involved. See Figure 7-13.

Figure 7-13: Coffered ceiling.

Ceiling Finishes

By definition, *ceiling finish* is the finished surface of a ceiling material. Paint is probably the most common and economical form of finish to be applied directly over plaster and drywall ceilings. Water-base latex paints are the most common type of ceiling paint. This type usually has a flat finish and an eggshell color to reflect light with relatively low glare.

Most solvent-base paints contain alkyd binders (used to bind the pigments together). This type of paint, however, is not appropriate for application over lead- or zinc-base paints. An undesirable chemical reaction between the two may take place, causing blistering of the painted surface.

Water-base paints have binders that are either alkyd or acrylic. Either one can be used as a ceiling finish. Acrylic coatings have the best service life under exterior conditions.

Paint color is a quality of the pigment used. Pigment composition may also affect paint quality. For the home inspector, the inspection of interior paint quality will seldom be a major factor. However, if the painted surface looks streaked or nonuniform, the painter probably failed to apply two coats. If this streaked condition exists, and two coats of paint were applied, the paint was probably of poor quality or else the work was not adequate.

Wood Finishes

Wood ceilings, such as cedar, need to have a finish that will seal the wood surface. A clear finish such as varnish is commonly used. An uncoated finish may tend to look dusty. A smooth, shiny surface is necessary for good uniform lighting through the room or area.

Varnishes are available to give either a glossy or low-luster finish. Both types are composed of various drying oils, resins, and solvents. Shellac, a resin-base varnish, is generally used to seal porous woods underneath a more durable top-coat varnish. Phenolic resins are durable and water resistant. Spar varnish, made with phenolic resins, is very good for outdoor applications including nautical vessels. Silicone treatments of unpainted masonry surfaces are quite common and consist of silicone resins and mineral spirits.

Interior Finishes

Urethane generally falls under the category of varnish. It is a clear, solvent-base coating that is often used for floor surfaces because of its durability. It is also a good choice for a clear coat on wood ceilings. Faux finishing is the name given today to hand-wrought or antique finishes that give various textures and artistic looks to a surface. This type of finish is generally associated with furniture, floors, and walls, but has also been employed on ceilings of many traditional and colonial homes. Marbleizing is a faux finish that imitates the look of marble. Various paint colors are blended by brushing them together on a surface. Wood graining is another faux-style finish that is more or less an artistic endeavor to simulate the grain of wood by means of a squeegee tool that is dragged through wet paint. Most faux finishes are best achieved by using oil-base paints. Washing is a finish that simulates finishes found in older homes. This technique is usually accomplished with water-base paint. A base coat is first applied. After this initial coat is dry, a second, extremely thin, coat of a slightly darker color mixture is applied so that the first coat shows through. This is accomplished by brushing or sponging so as to streak or texturize the finish coat. Experts are able to achieve very good results with these finishes if applied to appropriate settings.

A ceiling inspection can tell the home inspector quite a lot about the entire home, not just the ceiling itself. Therefore, the ceiling inspection should be thorough, following the recommendation outlined, along with the items on the ceiling checklist at the end of this chapter.

Plastered Ceilings

Old plastered ceilings and walls will almost always have cracks somewhere along their surface. Some of the reasons for these cracks are structural and settlement stresses, changes in humidity, and houses left vacant and unheated for several seasons.

Plastered ceilings will usually crack at the point of greatest deflection, or sag. If the crack is small, there may be nothing to worry about except for the unsightliness of the crack itself. In general, if the ceiling deflection is less than the length of the structural member in inches divided by 360, there will be no visual objection. For example, if the ceiling joists span 12' between wall partitions, then the allowable deflection is 0.4 inches.

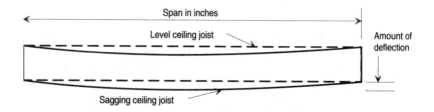

Figure 7-14: Ceiling deflection should not exceed the length or span (in inches) divided by 360.

$$\frac{144''}{360} = 0.4 \ inch$$

In other words, the ceiling can have a deflection of 0.4" ($1/10$" less than $1/2$") without being noticed. Any greater deflection, however, may be noticeable. A properly designed ceiling structure with a plaster finish will limit deflection to the length divided by 420, or an even larger number, to be on the safe side. Using the same span as in the example above, this would mean,

$$\frac{144''}{420} = 0.34 \ inch$$

or a little more than $1/4$ inch. These higher numbers are used not only to eliminate deflection but also to ensure that the plaster will not crack to an unsightly extent. See Figure 7-14.

Cracks in the ceiling do not necessarily mean structural failure. Joists may simply sag too much. Gypsum board ceilings may also show hairline cracks due to a poor joint compound or poor application of the compound. Drywall joints not taped properly or not covered with sufficient compound are likely candidates for hairline cracks.

Gypsum board joint compound is different today from what it was 25-30 years ago. The older compounds dried extremely hard and brittle. Today's compound is premixed and stays relatively soft, allowing it to absorb stress better than its earlier counterpart.

Wallpapered ceilings were common only a few decades ago. Cracks that occur in the ceiling plaster will usually show through the paper. Wallpaper usually doesn't split at these cracks; rather, it crinkles.

Papered ceilings will show the results of leaks by leaving brown stains. Such leaks will cause plaster to blister, bulge, and eventually become loose and drop off in chunks. All such defects should be noted by the home inspector.

Drywall Ceilings

Drywall or gypsum board will also stain and warp when it becomes wet, so look for such stains and bulges in drywall ceilings. If any are found, they should appear in your inspection report.

Most gypsum board ceilings are painted a flat finish that is usually an eggshell color to in crease the brightness of the room. A flat finish also tends to hide imperfections in the surface better than a gloss finish. In fact, it is often desirable to match the paint color with the color of the joint compound.

Inspecting drywall ceiling joints very carefully will tell you a lot about the skill of the painters.

If the painting was sloppy, you will be able to see brush or roller marks at the drywall joints where the paint did not fully cover portions of the joint compound.

Drywall joints tend to be the biggest problem with this type of finish. The edges of the gypsum board are beveled so that compound can be applied flush with the adjacent surface. If you can spot joints in the painted gypsum board, then the finishers did not spend enough time in sanding or did not apply the compound professionally. It will be next to impossible to find the joints in drywall ceilings when the jobs are finished carefully and properly.

Many ceilings are given a stippled or swirled finish using the joint compound. The inspector should check this type of ceiling finish out very carefully because one of the benefits of this finish is that it hides a multitude of ceiling deficiencies. A sloppy gypsum board ceiling finish can be hidden this way, making small cracks even harder to detect. The trick is to spot flaws such as sagging ceiling joists or other structural stress or settlement problems, then look closely for cracks in these areas.

Always carefully check ceilings above showers. Blistered paint, warping, mildew, etc., indicate that the bathroom does not have proper ventilation, or the improper type of gypsum board or finish was used.

Gypsum board in shower and bath areas should be the "green" moisture-resistant type.

Suspended Ceilings

From a home inspector's point of view, suspended ceilings are a convenient way to inspect ceiling structures and any utilities that may be installed above the ceiling. The lay-in ceiling tiles can be lifted up at almost any location to reveal the ceiling cavity above. If insulation has been laid on top of the tiles, inspect around all lighting fixtures, especially recessed lighting fixtures, to see that the insulation does not come in contact with them where dangerous heat buildup may occur.

If sound control is important from one room to another, soft lay-in ceiling tiles will do little to reduce noise. Sound-absorbing, acoustical tile should be used instead.

When inspecting suspended ceilings, check to make sure the ceiling tiles fit well and lay snug in the grid. Open and close a door or window in the area of the suspended ceiling. Does the grid rattle when doing so? Suspended ceilings should be hung from heavy gauge wires that are attached securely to the ceiling or floor structure above. If the wires have been stapled to the floor structure, then the ceiling probably isn't well secured.

Miscellaneous Ceilings

The home inspector should examine any exposed beams carefully. Exposed beams that are actually part of the ceiling structure will have an "honest" look in the framework, and both the future owners and the lending institution financing the home will have an interest in this fact, even if just for the aesthetic value. However, exposed beams may be of interest from another standpoint. If the bare underworks of the floor above is left exposed, there will be a lower insulation value between the two rooms; this will affect both the heat-loss and noise that is transmitted. Dirt and other debris from the attic space or floor above might find its way through the floorboards and settle in the living area of the home.

Exposed beams in newer homes are almost always fake. Fake beams can look nice and be quite satisfactory; however, they have absolutely no structural or support value. Such beams should never be used to

Interior Finishes

support even a small lighting fixture. Some of the newer styrofoam fake beams look like real wood, but a tap of the finger will identify such beams immediately. If styrofoam beams are exposed to fire they will give off toxic fumes.

Even real wood beams, used for decorative purposes, may have some limitations. Proper attachment of the wood to the building structure may be difficult for the home inspector to determine. These beams have been known to sag if heavy lighting fixtures or ceiling fans are attached to them. Some have even come loose and fallen to the floor. If at all possible, try to determine how such beams are secured and if found to be inadequate, show this on your inspection report.

Cathedral ceilings should also get a thorough inspection. One of the major problems with this type of ceiling is the tendency for the tops of the walls to push out from the force exerted by the sloping joists. Sometimes this is difficult to see, but if you look at the supporting walls from the outside, there will usually be a roof displacement that will show up in the exterior wall. This displacement occurs because the top of the exterior wall is unbraced . It is held in place only by the sloping rafters. Use a level to carefully check the plumbness of the exterior walls. Be sure to check both sides because if the wall has been brick veneered, the ceiling may have pushed the walls out of plumb prior to installing the brick. If this is the case, the exterior wall will not be as noticeable if the brick themselves were laid plumb. Therefore, check the plumbness of the interior walls if you suspect such a problem.

The home inspector should be particularly alert if a large room has a cathedral ceiling, such as over an enclosed swimming pool, especially if there are no horizontal ceiling struts or cables and no horizontal collar ties at the apex of the ceiling.

Generally, the best design of an unbraced cathedral ceiling is one that employs horizontal ties of some type. Installing such ties is also the best corrective measure.

Basement Ceilings

In basements of both old and new homes, the home inspector can usually find a part of the ceiling that is unfinished. This is where the home inspector can examine the home systems such as ventilating ductwork, electrical wiring, and plumbing pipes. In doing so, note whether the cold water lines are insulated. A basement may have a high relative humidity which will cause water to condense on the pipes and drip. If installed above a finished ceiling, this condensation will drip onto the backside of the ceiling causing stains, peeling, and/or blistering of the finish.

Ductwork provides a means of moving air from the heating and cooling system to the rest of the house. Note if such ductwork is insulated. In older homes, this ductwork is normally round and will be insulated with asbestos. Therefore, wear a face mask when doing the examination. Asbestos is an excellent insulation, suited for insulating heating ductwork. Unfortunately, it has been found to be toxic and is no longer manufactured.

If the ceiling joists are accessible, check them very carefully for wood decay and mildew. Wood decay may occur only in localized areas. The wood surface will be soft and can be scrapped out with a pocket knife. Sound wood cannot be scraped away with the knife; rather, it will split. Wood decay, sometimes called "dry rot," can result in structural failure. Corrective measures include splicing the affected joist by bolting two pieces of timber onto both sides of the joist at the weak section. A better way is to remove the affected joist or beam completed and replace it with a new one. Another solution is to add a column and a column top plate under the weak spot.

Ceiling Insulation

The most important ceilings that require insulation are those that are directly below the roof. If the attic is unused and contains nothing that might be affected by extreme temperatures (water pipes, etc.), then heat loss should be controlled at the ceiling of the room(s) immediately under the attic space. During cold weather, a warm, humid interior will transmit water vapor to the cooler attic, resulting in condensation on the inside of the attic sheathing, especially on the underside of tin roofs. Check for vapor damage to the roof sheathing. The attic requires proper ventilation to remove this moisture, and if improper ventilation is discovered, your report should recommend that the situation be corrected.

The ceiling inspection checklist in Figure 7-15, will help the home inspector during his or her inspection of ceilings in residential construction.

Interior Finishes 149

INSPECTION CHECKLIST — CEILINGS				
	Circle one of the following:			
Major ceiling material:	Plaster	Drywall	Tile	Other
Ceiling types:	Secured to framing above	Suspended ceiling	Cathedral ceiling	Exposed beams
Other:				

		Yes	No
1.	Are there any cracks or chunks missing from ceiling?		
2.	Are there any water stains on the ceiling?		
3.	Are popped nails visible?		
4.	Does the ceiling sag?		
5.	Are exposed beams real?		
6.	Are any exposed beams fake?		
7.	Are any lighting fixtures, ceiling fans, etc. secured to any fake beams?		
8.	Is all wallboard secure?		
9.	Are ceilings in good shape; that is, no peeling paint or faded colors.		
10.	Are there any bubbles in the ceiling surfaces?		
11.	If visible in the basement area, are all cold-water pipes insulated?		
12.	Is the house free of asbestos insulation?		
13.	Are ceilings free of holes?		
Additional features:			
Inadequacies:			
Repairs needed:			
Quality of installation: Good Fair Poor			

Figure 7-15: Inspection checklist for ceilings.

WINDOWS AND DOORS

Windows and doors should be checked during the exterior inspection, but they should also be examined during the interior inspection. Sometimes the windows and doors will look fine from the outside of the home, but a closer examination on the inside will reveal many discrepancies. For example, children may have abused the interior doors — banging toys or other sharp objects against them. Pets may have left scratch marks trying to get in or out of the home. Before getting into the actual inspection, however, let's find out something about windows and doors.

Interior Doors

The main differences between interior and exterior doors are in width and height. An exterior door is solid, and so, heavier, and is about three feet wide. Interior doors are lighter (most interior doors today are flush, hollow core) and are narrower. Hardware is different in that exterior doors can be locked and opened from either side. Interior doors are locked and opened from the inside or a room in most cases, unless some special reason dictates that it can be locked to prevent entrance except with a key. Examples could be home workshops, home offices, a room or closet containing firearms, etc.

Since interior doors are hollow core they are usually hung on two hinges. Heavier exterior doors require three. Conventional doors require a header above the door during framing, while the newer full-height doors do not require the header. In recent years, doors are purchased in an assembly that includes the jambs and the door already hinged in place. All that is required is to set it in place, make sure it's level, and then secure it.

Doors are sometimes used in residential construction that go from floor to ceiling. This gives the advantage of more headroom and easier framing and doesn't chop up the wall surface.

Sliding doors are especially suitable where swing space is to be conserved. A hinged door requires an area on the hinge side so the door can swing open. The wall against which it opens must be clear for at least the width of the door. A bumper should also be provided to prevent marring the door and the wall when they make contact. None of this applies to a sliding door.

Chapter 8
Exterior Finishes

Siding materials are used on the outside of the home not only to create a finished appearance but also to protect the interior against the elements. Siding materials can be made of solid boards, wood shingles or shakes, plywood panels, or hardboard panels treated with grooves to create the appearance of individual boards vertically placed to accentuate height.

There are also nonwood siding materials such as asbestos-cement shingles, aluminum shingles and vinyl that are made into long siding strips to create the appearance of wood boards.

Other siding materials include brick veneer, stone veneer, stucco finish, and other masonry products.

This chapter covers all of the popular outside finishes for residential applications throughout the United States — including windows and exterior doors. You will also be shown how to inspect other outside parts of the home; that is, soffits, fascias, overhangs, guttering, porches and stoops.

WOOD SIDING

Wood siding is probably the most often used material on houses, although many types of imitation wood products have been used over the past decade or so. The most common woods used for siding include:

Western white pine	Western hemlock
Sugar pine	Ponderosa pine
Eastern white pine	Spruce
Cedar	Poplar
Cypress	Douglas fir
Redwood	Yellow pine

Bevel Siding

Bevel siding material is furnished in 4-, 5-, and 6-inch widths with $7/16$-inch butts. In all siding the top edge is usually $3/16$ inch thick for all sizes. This material is furnished in random lengths varying from 4 feet to 16 feet. See Figure 8-1.

The minimum lap for bevel siding should not be less than 1 inch. Installation begins with the bottom course and is normally blocked out with a starting strip the same thickness at the top of the siding board. Each course overlaps the lower course. Siding should be nailed to each stud centered on 16 inches.

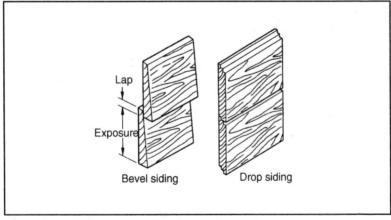

Figure 8-1: Horizontal siding.

Exterior Finishes

Seven-penny or eight-penny nails (2¼ inches and 2½ inches long) are the types most commonly used for securing bevel siding to the wall sheathing of a house. Where fiberboard, chipboard, or gypsum sheathing is used, the ten-penny nail gives better penetration into the stud. For ½-inch-thick siding, nails may be ¼ inch shorter than those used for ¾-inch siding.

In nailing siding boards, the nail should be placed far enough up from the butt to miss the top of the lower siding course. Siding board-end-butt-joints should be made over a stud for proper end nailing, and butts should be staggered between courses.

Drop Siding

This material is usually ¾ inch thick and is made in a variety of patterns with either matched or shiplap edges. Shiplap exterior wall covering boards have rabbeted edges that allow for a fitted overlap between edges.

Drop siding can be nailed directly to the studs — serving both as sheathing and exterior wall covering — in mild climates for garages, sheds, and farm structures. All homes, however, should have the boards nailed over some type of sheathing.

One of the most common problems with this type of siding is water working its way through the joints that sometimes cause paint failure and wood decay. These conditions, however, are practically eliminated on houses with large roof overhangs.

Miscellaneous Wood Siding

Horizontal wood sidings shown in Figure 8-2 (on the next page) are: plain bevel and bungalow; rabbeted bevel and bungalow; Anzac; shiplap and rustic; tongue and groove, and board and batten.

Vertical siding is commonly used on gable ends of houses, over entrances, and also for large wall areas. See Figure 8-3 on the next page.

The various types of vertical siding are plain-surfaced matched boards, patterned matched boards, or square-edge boards covered at the joint with a batten strip. Matched vertical siding should be not more

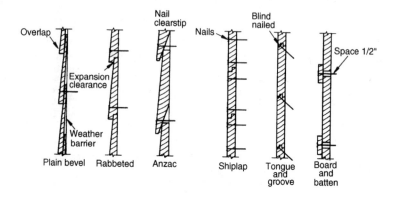

Figure 8-2: Various types of horizontal wood siding.

than 8 inches wide and should be fastened with two eight-penny nails spaced not more than 4 feet apart.

Hardboard Siding: Hardboard siding may be used directly over sheathing, or it can be applied directly to the studs without the use of sheathing. The material is made in panels, 4 feet wide by 8 or 9 feet long and $7/16$ inch thick. It is also shaped into siding boards, 12 inches wide by 16 feet long and $7/16$ inch to $5/8$ inch thick. See Figure 8-4.

The panels and siding boards are prefinished before they arrive at the job site. Surfaces are made to look like a skip-trowel texture, giving

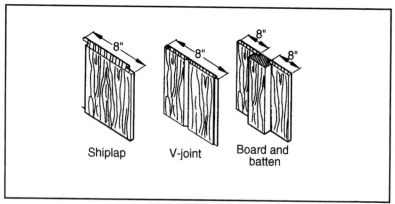

Figure 8-3: Some types of vertical wood siding.

Exterior Finishes

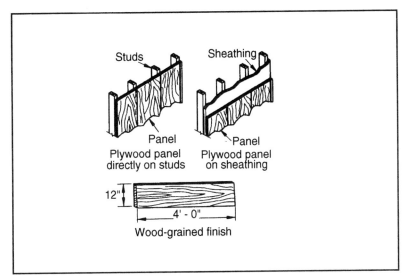

Figure 8-4: Horizontal siding boards.

the boards a cement-stucco appearance. Some types have small indentations to give them a hammered appearance. Some siding boards may have a rough cedar texture, giving them the appearance of wood or shingles. Others have a deep embossed bark-like surface texture.

Plywood Siding: Plywood siding panels are made in 4 feet x 8 feet panels. Thicknesses are usually $3/8$ inch to $5/8$ inch. The $3/8$-inch panel is three-ply, while the $5/8$-inch panel is five-ply. Panels are constructed to form $1/8$-inch or $3/8$-inch-wide channel grooves.

A plywood panel may have three vertical grooves $5/32$ inch or $3/8$ inch wide by $3/32$ inch or $1/4$ inch deep. This gives the appearance of three individual boards, rather than just one 4 x 8 panel.

Board and Batten Siding: Board and batten siding designs are used vertically on building walls. Wide boards are first nailed to the wall sheathing (in a vertical position) and then narrower strips of wood are nailed over the joints of the wider boards. These strips are called battens. See Figure 8-5 on the next page.

Batten and board siding application is similar to board and batten siding except that the wide boards are nailed over the batten boards. Board and board siding employ boards of the same width. All three types are shown in Figure 8-5 on the next page.

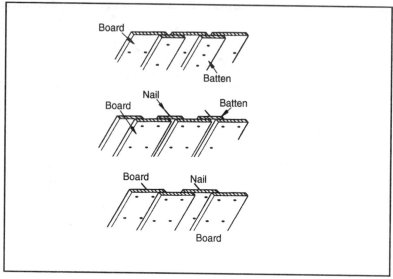

Figure 8-5: Vertical board siding.

Shake And Shingle Applications

Maximum recommended weather exposure with single course wall construction is 8½ inches for 18-inch shakes and 11½ inches for 24-inch shakes. Nailing is normally concealed in single-course applications; that is, nailing points are slightly above the butt line of the course to follow. Double-course applications require an underlayment of shakes or regular cedar shingles. Weather exposures up to 14 inches are permissible with 18-inch resawn shakes, and 20 inches with 24-inch resawn or taper-split shakes. If straight split shakes are used, the double course exposure may be 16 inches for 18-inch shakes and 22 inches for 24-inch shakes. Butt nailing of shakes is required with double course application. The nail heads should not be driven into the shake surface. Doing so will weaken the shake and limit its ability to shed water.

The two methods of shingled sidewall application are single course and double course. In single coursing, shingles are applied the same way in roof construction, except that greater weather exposures are permitted. Shingle walls have two layers of shingles at every point.

Double coursing allows for the application of shingles at extended weather exposures over undercoursing grade shingles. Double

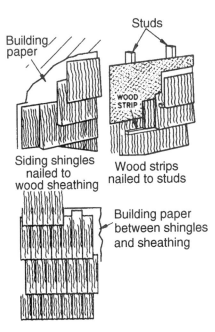

Figure 8-6: Single and double coursing shakes.

coursing gives deep, bold shadow lines. when double coursed, a shingle wall should be tripled at the foundation line by using a double underlay. When the wall is single coursed, the shingles should be doubled at the foundation line.

Figure 8-6 illustrates siding shingles nailed to wood sheathing with a felt building paper underlayment (between shingles and sheathing), as well as wood strips nailed to studs for lining-up shingle courses. The outer course extends ½ inch lower than the undercourse. The shingles are nailed 2 inches above the shingle butts.

When nailing for double coursing, each outer course shingle should be secured with two 5d (five-penny) (1¾ inches) small head rust-resistant nails driven about 2 inches above the butts, ¾ inches in from each side. Additional nails spaced above 4 inches apart should be applied across the face of the shingle. Single coursing involves the same number of nails but they can be shorter 3d (1¼ inches) and should be driven not more than 1 inch above the butt line of the next course. The wood should not be crushed when the nails are driven.

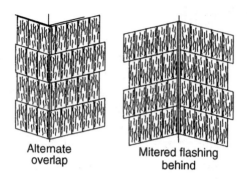

Figure 8-7: Outside and inside corners of wood shingles.

Corners

Outside corners should be constructed with an alternate overlap of shingles in successive courses. Inside corners should be mitered over a metal flashing, or they may be made by nailing 2-inch square strips in the corners, after which the shingles of each course are jointed to the strip as shown in Figure 8-7.

MISCELLANEOUS WALL SIDING

Asbestos-Cement Siding

Asbestos-cement siding sheets are usually 42 inches wide and have 10 corrugations per width of the sheet. The thickness is approximately $3/8$ inch. The sheets come in lengths ranging from 6 inches to 12 feet, in multiples of 6 inches. Asbestos-cement sheets, as the name implies, are made of cement and asbestos fibers. The natural color is a gray-white.

Asbestos-cement panels are no longer used on new construction as asbestos is thought to be harmful to people's health. However, the home inspector will more than likely find such siding on some existing homes. They are particularly plentiful in seacoast areas, since they are not affected by salt air.

Exterior Finishes

Aluminum Siding

Aluminum siding comes in vertical and horizontal designs. The horizontal siding is made with a weather exposure of 4 inches, 5 inches, and 8 inches, the 8-inch type being the most common. These sidings have been installed without backing, polystyrene backed, and fiberboard backed. The backings provide extra insulation along with rigidity. Furthermore, this backing helps to cut down on the noise during a rain storm, or "popping" during a temperature change.

The thickness of the aluminum is .024 inch. The polystyrene backed siding is $7/8$ inch thick, and fiberboard backed siding is $3/8$ inch thick. The siding lengths are 12 feet, six inches.

The aluminum surfaces are finished in a variety of colors and patterns. Some even have a wood grain pattern with colors of white, tan, sandstone, and fern green.

Vertical aluminum siding is usually 12 inches wide and is made to give an appearance of two individual 6-inch vertical panels. Some panels are available with raised batten strips. Colors are the same as those on horizontal siding. Exposure widths are 12 inches and the panel length is 10 feet.

Aluminum siding in the form of imitation cedar shakes is also available and manufactures claim that such siding requires 75% less nailing than conventional shingles. Furthermore, they will not burn or support combustion, rot, warp, split, crawl, deteriorate, or need replacement like conventional wood shingles. Properly installed, aluminum cedar shakes will withstand winds up to 110 mph and an uplift of 69 pounds per square foot. The air space between the shakes and the wall sheathing also helps insulate the entire home against the elements.

Aluminum cedar shakes are easily installed across wall areas where panels work well around windows and door openings. A wide variety of trim accessories helps to speed up and simplify the installation.

Vinyl Siding

Vinyl siding all but replaced aluminum siding for homes during the 1990s, to provide a maintenance-free wall covering; the trend continues. In general, vinyl (pigmented polyvinyl chloride) siding is used in the same way as described for aluminum siding; that is, to be applied directly to the wall sheathing or else to cover existing wall coverings that have deteriorated. All vinyl siding should have some type of backing. Alone, it has no structural strength as wood siding does, and if no backing is present, it is easily cracked, and must be replaced; temporary repairs are not recommended for this type of siding. Furthermore, if the sheathing is weak underneath — due to wood rot, decay, etc. — the vinyl siding offers no additional strength and the entire home may be severely weakened.

Properly installed vinyl siding will last for years, but an improperly installed job will start causing problems immediately, such as having planks blown off in a wind storm, cracking, leaking, and the like. Like other plastics, it is also subject to fading from sunlight.

Log Siding

In recent years, prefabricated log homes are becoming extremely popular throughout the United States — mainly for vacation or second homes. In fact, hundreds of companies offering these homes sprung up in the 1980s — from Maine to California — and most have enjoyed success.

Log homes are structurally sound, and most manufacturers guarantee their materials and workmanship for a period of ten years or more; that is, any defects that occur due to materials or workmanship on the part of the manufacturer will be repaired or replaced free of charge up to 10 years from the date of purchase.

Log home have natural wood insulating qualities. In just once cubic inch of wood, there are about 3 million hollow cells. The insulative qualities of four inches of wood are equal to a concrete wall 5 feet thick. Wood has the many qualities of incredible strength, durability, and ease of maintenance.

In general, there are two basic types of log construction: full log construction as shown in Figure 8-8, and insulated log construction as whereas logs are used as a veneer over conventional framing. In most cases, the logs are treated at the mill with a colorless, non-toxic sealant to protect them from mildew and mold. After the home is constructed, a more permanent sealant and stain should be applied to protect the surface.

Exterior Finishes

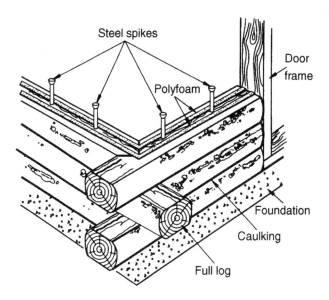

Figure 8-8: Typical features of full-log construction.

MASONRY FACINGS

When a mortar mix containing portland cement is used as a coating for interior walls it is called *cement plaster*. When an identical mix is used for exterior work it is called *stucco*. There is absolutely no difference between cement plaster and stucco as to its content and application. The difference is in name only.

Stucco is an ideal coating in areas subject to extremely hard use and abuse, or where moisture is excessive. It can be applied directly over concrete block, brick, or poured concrete walls if such surfaces are unpainted, clean and rough.

Stucco applied over wood siding or sheathing requires some preparation of the base. In general, a 15 pound waterproof felt is applied to the surface using galvanized or other rustproof nails. This felt is lapped 4 inches on horizontal joints and 6 inches on vertical joints. The paper is doubled at all corners. Caulk seams must be applied around doors and windows and these seams should then be further protected with felt so that the door and window frames don't absorb moisture and swell. When inspecting buildings with a stucco finish, always check these seams around doors and windows to make sure they are sealed tight. Otherwise, structural damage could be present.

Rustproof metal lath, using special rustproof lath nails is then applied over the felt. The lath nails are designed so that the lath will be offset about $\frac{1}{4}$ inch from the wall, which allows the stucco to be squeezed behind the lath and form a bond. It is also important to look for rustproof flashing at areas where it is needed to prevent water from getting behind the stucco.

The standard mixture for stucco is 3 parts sand (or other aggregate) to 1 part portland cement. If you should observe cracks in a stucco-finished building, this is an indication that the stucco mixture was too rich (too much cement) which usually results in shrinkage cracks. There's more thorough information on inspecting stucco finishes later on in this chapter, including how to report on same.

Surface Coatings

Paint, as we know it today, originally consisted of a natural oil such as linseed oil, that served as a base to which finely ground solid particles of white lead, called pigments, were added. After proper mixing, the white lead particles were held in suspension in the base — sometimes called "vehicle." Paint could be thinned by adding a solvent thinner, and it could be removed by applying a solvent such as turpentine. White paint was colored by adding pigments.

Children have been poisoned by eating old flaked-off paint with lead bases. Consequently, there are laws prohibiting the use of lead-base paints in residential structures. Low-lead or lead-free latex paints are now used almost exclusively for interior applications. High-lead paints can be used as outside paint, however, because lead poisoning is less likely to occur, as outside paint is not likely to be eaten by children. Those homes with metal roofs, however, that utilize a cistern for their water supply should not use a lead-base paint on the roof.

It is highly likely that lead-base paints will be disregarded altogether in all but industrial applications. Many painters have suffered lead poisoning that eventually caused kidney failure and death.

Once applied by either brush, spray, or roller, the paint forms a protective film when dry. And this protective coating is very important to the life of exterior finishes. In their unprotected state, steel rusts and

Exterior Finishes

wood weathers. Consequently, the thin protective coating of paint (or other finish) is highly important for all wall coverings with the possible exception of aluminum and vinyl.

To offer the best protection, paint must be applied in three or more coats: a prime coat (called "primer"), an undercoat, and a finish coat.

A complete change in paint and painting procedures occurred with the development of synthetics and a variety of white paint solids. This change in paints and techniques is still continuing.

PAINT BASES

Newer bases, such as resins or natural oils, form the paint film. The base holds together the white paint solids and causes the paint film to adhere to the surface. The base also supplies the protective and durable qualities to the film. All of these play a very important role in the hiding qualities of paint.

INSPECTING THE HOME'S EXTERIOR

During any home inspection, the inspector should observe and report on the following. The form at the end of this chapter will facilitate doing so.

- Exterior wall coverings and trim.

- Outside windows and doors.

- Porches, areaways, steps, stoops, balconies and decks.

- Eaves, soffits and facias, including the attached guttering and downspouts.

- Any storm or screen windows and doors that are in place.

Wood Siding

Upon approaching the home, and at some distance from the home, observe the building lines to see that they are straight and plum. Any lines that are not straight and plum should be reported, as this could mean structural weaknesses.

Once you are closer to the home, examine the wall covering carefully. Wood siding should be tightly secured to the framing, with no splits, warps, or other conditions that could allow water to seep between the outside covering and the sheathing. This condition can cause wood rot. Any suspicious areas encountered should be probed with either an ice pick or screwdriver. If wood rot or decay is evident, this should be noted on your report.

If the paint finish shows excessive bubbling, cracking or peeling, it may mean that the house has insufficient vapor barrier protection; that is, inadequate insulation, or insulation that does not have vapor-barrier protection. A vapor barrier is designed to stop the moisture in a house from penetrating through the walls where it may condense when it contacts cold outside air.

You will also want to note any missing siding which will definitely allow unwanted moisture, insects, and other undesirables into the house, weakening the wall structure and causing damage to the finished interior. Such conditions are also prone to having more covering removed during wind storms, or as moisture penetrates the siding, the fasteners securing the siding will become loose. So carefully look for such conditions on every inspection.

Masonry Siding

Brick is probably the sturdiest building material for residential walls, and such coverings are still in existence after being subjected to the elements for hundreds of years. The first thing that you want to determine is if the walls are brick veneer or solid masonry. Brick veneer is a course of brick used only as a facing material without utilizing its load bearing properties, while solid masonry utilizes the bricks and/or cement blocks to carry the roof load.

The major problem that occurs with brick veneer is the mortar that holds the bricks together. Mortar can dry, crack and crumble through the combined effect of the elements, and if left unchecked, can eventually lead to a high repair cost. If crumbly mortar is not replaced at an early stage, it can lead to the problems described for wood siding — allowing moisture to enter the inside of the home that will eventually cause both structural and interior damage. Improperly mixed mortar is another cause of defective joints. For example, mortar with too much sand may be too weak; mortar with too little sand may be too brittle. Joints should be tooled so that water is able to drain off the wall, not be trapped in the joints.

Also look for cracks in the mortar joints and in the bricks or cement blocks themselves. Cracks in masonry walls are usually caused by settling of the foundation, but cracks can also occur if iron lintels begin to corrode. Rust occupies many times the volume of the iron it displaces. A rusting lintel is capable of lifting several dozen courses of brick above it. If the rusting takes place too rapidly, the masonry will crack. If at a slower rate of oxidation, perhaps only a joint or two will crack.

When checking for cracks, look at the tops of all window and door openings. These places are where masonry cracks will usually occur first — usually due to using lintels that are too short. When short lintels are used, the bricks at the end of the lintels cannot support the required weight and cracks will develop. The same situation will occur when wall openings are placed too close to wall corners.

If the surfaces of brick have a mealy or powdery look, this indicates that moisture has seeped through behind the brick. When moisture is drawn through the brick to the outside to evaporate, it dissolves salts that remain on the brick's surface. If this condition occurs on any wall that is not relatively new, look for the source of the moisture. It can seep through cracks in the mortar, seep down from above, or seep up from below in a damp basement area.

A stucco finish can peel and crack and if not attended to immediately, repairs will require more than an occasional paint job. Therefore, carefully check all stucco surfaces for settlement cracks that might indicate a structural problem. Look for off-color patches in the stucco wall — possibly suggesting problem areas that may or may not have cured properly. Such patches may also indicate an excessive amount of moisture seeping into these areas.

Aluminum/Vinyl Siding

Aluminum siding is almost maintenance-free — requiring only an occasional wash down with a garden hose to keep the home looking good. Vinyl siding has many of these same features. However, both are subject to noise during rains and if one of the fasteners should become loose, you can bet that this piece of siding is not going to be on the house very long. Once one piece is missing, winds and the elements will tend to rot the adjacent wood framing, causing other fasteners to work loose, not to mention the damage to the framing and interior finishes that may follow. Therefore, you should check all aluminum and vinyl planks to make sure they are firmly secured. Also tap on the siding to see if it is hollow or backed with some support. Make notes of your findings.

When inspecting aluminum siding, always check for dents and scratches; check for brittle areas and cracks in vinyl siding. Both of these defects should appear on your report.

Windows And Doors

Door must be level and plumb to operate properly. Open and close all outside doors several times. Does it scrape or bind? Is it hard to open? If so, something is out of line. Either the door is improperly hung, the door frame is out of line, or the floor is not level. The latter case will occur the most often when foundations settle. Unless the door problem is caused by poor workmanship (improper handing), tight or binding doors can usually be corrected easily.

All sides of exterior doors should be weather-stripped to prevent cold drafts from entering the home, especially at the bottom of the doors. Besides being uncomfortable, such drafts waste heating fuel. Check for weather-stripping at the top, bottom, and both sides of doors. Any found missing, note this on your report.

A squeaking door may need only a few drops of oil on the hinges, but then again this squeak may be caused by a warped door that places a strain on the door frame. Severely warped doors usually must be replaced. Any amount of planing or weather-stripping will not solve the problem. Of course, all replacement doors must be designed for exterior use, and have high insulating qualities.

Don't overlook the door frame and trim. Note any cracks or deteriorating materials. Probe with your ice pick if wood rot or termites are suspected. Rotting or decaying door frames or trim could be the reason why a door closes hard.

Finally, note the condition of the finish on all outside doors (both inside and out). Is the paint peeling or cracking? Are the doors sound?

Then check the locks and latches on all doors to be sure they open, close, and lock easily.

To perform a thorough inspection job, open and close all windows in the house to determine how well they operate and fit in their openings. A window that is out-of-plumb will bind or close hard, or be difficult to open. Again, house settlement can be the cause of out-of-plumb windows although an improper installation is not out of the question.

When performing the opening-and-closing test, also check the window locks. Make sure they work easily and pull the sashes together snugly. Make note of any broken window locks. An improperly installed window can put undue stresses on the lock, and will eventually break them.

Loose fitting windows may require weather-stripping to prevent drafts. If weather-stripping is installed, however, make sure the windows still operate smoothly. While performing this inspection, also check for loose or cracked window panes. Such conditions can be the cause of drafts and allow moisture to seep into the home's interior. Loose panes can usually be cured by resetting with glazing compound, but cracked or broken panes will obviously have to be replaced.

Window frames and trim should be inspected as described for exterior doors. Check for water stains on the sills which is an almost sure sign of leaking windows.

If storm windows or screens are in place, check these at the same time. However, you are not required to check them if they are stored, say, in the basement or garage; only if they are in place.

Outside Trim

When inspecting the siding, also check the eaves, fascias, soffits, and all outside trim, including any shutters. Boxed in soffits sometimes pull loose encouraging birds and insects to enter these areas for nesting. Any loose boards in any of these areas should be secured snugly.

Faulty guttering can cause rotting and decay in the soffits and any suspicious areas should be probed with an ice pick or screwdriver blade. Indicate what you find on your report. Also note the condition of the finishes in these areas; that is, condition of paint or other covering. Are there any cracks or discolored surfaces?

Outside Coverings

All painted surfaces begin to deteriorate immediately after installed due to exposure to ultra violet radiation and the elements. Sun screens, anti-oxidants and inert pigments in the paint slow the process, but eventually the vehicle breaks down and the surface becomes dull.

When inspecting outside wall coverings, look for surface deformation, staining, cracking, separation of layers, and separation of the coating from the base. If any of these defects are found, you should also try to determine the reason.

Deformation: This condition is usually caused by the paint be improperly applied; that is, in direct sun light that causes the solvents in the covering to evaporate too quickly. The same condition occurs, however, if the coat of paint was applied too heavily. The result is wrinkles in the finish coat, changing the reflective value of the paint and also tending to get dirty quicker.

Stains: The most common stains found in exterior wood siding will be from rusted nails, but there are numerous other stains: sap or resin deposits, mildew, soil stains, and a host of others. None of these usually seriously affect the home's overall condition; merely the appearance. Such stains should, however, be noted on your report.

Cracks: Hard, brittle paint is usually the culprit. Old, hard paint is no longer able to respond to the dimensional changes in layers of old finish or the siding underneath induced by changes in temperature and humidity.

Peeling: Paint peeling is due to poor adhesion, causing the layers to separate. A new coat of paint may hide the defect temporally, but if the new coat does not have a way to bite into the old finish, it will soon peel also. The only proper way to correct such a condition is to scrap or otherwise remove the poor layers underneath, properly prime the wood or metal, then apply a base coat. Finally, apply the finish coat at the correct temperature and out of direct sunlight.

INSPECTION CHECKLIST

Many experienced home inspectors can tell a lot about the entire home by the condition of the home's exterior. For this reason, the exterior is usually the place to begin with any home inspection, and it should be thorough.

Exterior Finishes

The checklist in Figure 8-9 will assist you in checking all exterior elements of the home.

	INSPECTION CHECKLIST — EXTERIOR FINISHES		
		Yes	No
1.	Is the house visually plumb and straight?		
2.	Is the siding securely attached?		
3.	Are there areas where siding is missing?		
4.	Are ther areas where moisture might penetrate due to splits, warping, etc.?		
5.	Does the exterior paint finish show signs of peeling or cracking?		
6.	Are mortar joints in good condition?		
7.	Are any mortar joints cracked?		
8.	Are there any cracks in the bricks or blocks?		
9.	Are there cracks above doors and windows?		
10.	Do door and window lintels have sufficient bearing length?		
11.	Are any settlement cracks visible?		
12.	Are there any cracks that water can penetrate?		
13.	Are there off-color patches indicating moisture penetration?		
14.	Do doors open, close, and lock easily?		
15.	Are all exterior door fully weather-stripped?		
16.	Do windows open, close, and lock easily?		
17.	Are any windows loose-fitting and drafty?		
18.	Do any windows need replacing?		
19.	Do any doors need replacing?		
20.	Are eaves, soffits and facias full covered and impenetrable?		

Figure 8-8: Inspection checklist for the home's exterior.

Chapter 9
Insects, Vermin and Decay

During any home inspection, you may find evidence of insects, vermin, and wood decay. Pests are certainly unwelcome in any home, but many of them can actually be destructive to the building's structure and/or the health of the occupants.

Detecting and reporting on any of these pests, however, must be done with caution. In many areas, only licensed pest-control experts or registered structural engineers/architects are authorized to inspect for wood damage in buildings. Consequently, before inspecting for, or reporting on, home pests, you should check all state laws and local ordinances.

If you suspect insects, vermin, or wood decay in a home, you should not withhold this fact from your client. Without going into details that may be illegal in your area, you could simply suggest that the home be inspected for insects, vermin, and/or decay by a licensed professional.

This chapter is designed to acquaint you with the various home pests, how to detect them, identify what you have found, and finally, how to write your report.

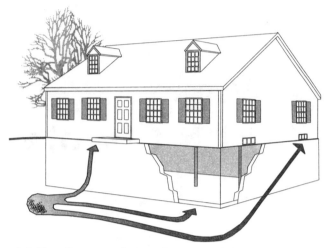

Figure 9-1: Termites nest close to homes and then tunnel their way to the house for food.

TERMITES

Nothing can be more alarming to most homeowners than the knowledge that termites are eating in the same house. There seems to be an immediate fear that the house is gong to come tumbling down at any minute. The truth is that there are few cases on record of any homes collapsing because of termite infestation.

That doesn't mean their presence should be ignored. When termites are discovered in a house, the inspector should recommend that a pest-control expert be hired. It's a time for action and not worry. This is the only way to get rid of termites and prevent their return.

Termites usually nest close to the house with tunnels to several spots along the foundation as shown in Figure 9-1. One of the best termite-prevention devices is a termite shield. A termite shield is merely a length of sheet metal installed on top of the foundation wall between the masonry and wooden members as shown in Figure 9-2. The bent edges act as a barrier and prevent the termites from climbing onto the wooden members.

How can you be sure the house is infested with termites? The most common variety of termite looks something like a flying ant to most people, but to an authority there is a decided difference. A flying ant

Insects, Vermin and Decay

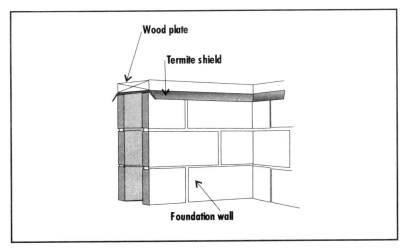

Figure 9-2: Termite shields help to prevent the entrance of termites into the wooden members. All new homes should have these shields installed. The absence of termite shields should be used in the inspection report.

has an hour-glass figure; a termite has a body that is thick from end to end. Actually, the chances of seeing a termite are rather remote, but if you should find a lot of silver-colored wings on the basement floor, or in a crawl space, sweep them into an envelope for an expert analysis.

The insect in Figure 9-3A is the one that does all the damage. Note that the wings are of even lengths compared to wings of unequal size on the ant (sometimes called "flying ant") in Figure 9-3B. You will

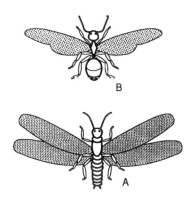

Figure 9-3: Comparison of winged termite and flying ant.

Figure 9-4: Worker termites (left) have straight bodies with no waistline as the ant on the right.

also notice the relatively straight body on the termite, while the ant's body is kind of wasp- or hour-glass shaped. Both insects, however, only have wings during the mating season. See Figure 9-4. Here both of the insects are shown without wings. Again, notice that worker termites (left) have straight bodies with no waistline while the ant (right) has an hour-glass figure.

Termites live on cellulose, so the first means to prevent termites is to get rid of any old wood, clothing, books, etc. that are in or on the ground near the house. They thrive on dampness, another of several reasons why crawl spaces should be well ventilated. Most varieties of termites build tunnels from the ground along foundation walls to the wood that becomes their food.

Pushing a pen knife blade into some of the wooden members of the house will determine whether the termites have done any serious damage. The blade will sink in easily if the insects have eaten away the inside.

If termites are found and identified, termite experts can rid the home of termites and this same firm can take steps to keep them away.

Ants

Several types of ants are prevalent in most parts of the United States. Of over 2,500 known species of ants, most varieties are more of a nuisance rather than a cause of damage to the home. Others, like the carpenter ant, can cause more damage to the building's structure than termites because they are capable of removing large amounts of wood in a short time. They will generally attack a soft spot in the wood and branch out into the harder, outer wood as the colony grows. Their activity may go undetected for some time because their tunnels are usually located in the center of structures and not on the surface.

Like most ants, carpenter ants remove materials from the nesting site as they build. Unlike termites, the carpenter ant does not actually eat the wood fiber (cellulose), but merely excavate within the wood. They will often remove sawdust and other materials left over from construction or from previous infestations. This inclination for cleanliness is also a good indication of a possible infestation. Piles of wood material resembling sawdust or pencil shavings, mixed with the remains of exoskeletons (body casings that are shed as the ant grows) and dead ants are clues that carpenter ants are at work.

The presence of the winged variety of the carpenter ant is another strong indication that the ants are living within the wooden structure. Winged ants indicate that the ants are breeding and spreading colonies.

It is possible to deter the ants from a structure by making entry difficult. Tree branches that touch the building should be trimmed away, denying the ants access from trees and bushes which may already be infested.

Carpenter ants can be eliminated with poisons, but steps must also be taken to make the nesting site unsuitable for future colonies. By repairing faulty plumbing and guttering located near the nesting site, the source of moisture is eliminated, making the site uninhabitable for carpenter ants. See Figure 9-5.

Figure 9-5: Carpenter ants can actually be more destructive to a building structure than can termites.

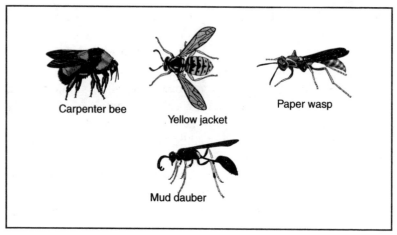

Figure 9-6: Flying insects found around the home.

Bees and Wasps

Figure 9-6 shows several types of flying insects that might be found during a home inspection. Like most ants, bees and wasps are considered more of a nuisance than a threat to the structure of the building. The painful stings delivered by many varieties are annoying at the least, but can be deadly to a person who is allergic to the venom injected with a sting.

One variety of bee can be a threat to the structure of a building, namely the carpenter bee. Like the carpenter ant, the carpenter bee makes tunnels in wood, leaving behind a small pile of sawdust to announce its presence. The holes are usually about ½" in diameter and found in untreated wood areas such as decks, woodpiles, etc. The bees resemble bumble bees except that they have shiny abdomens.

Roaches

Man has provided the roach with the ideal habitat: human dwellings. The roach fairs best when it is supplied with ample food and the temperature remains moderate. Modern homes provide the roach with everything it needs to survive; food, shelter from the elements, and a temperature that remains between 65 and 75°F all year long. Given

Insects, Vermin and Decay

Figure 9-7: The roach finds its ideal habitat in human dwellings.

these conditions, a roach population can thrive if allowed to exist unchecked. See Figure 9-7.

It is difficult to prevent the roach from entering the home, and even more difficult to eliminate them once they take a foothold. Roaches are attracted by scents emitted by decaying or fermenting organic matter, so it is no surprise that they are found in abundance in areas where food is abundant. They thrive in cities because of the amount of garbage that is easily accessible, and because of the variety of places they can hide when efforts are made to exterminate them. They are known to crawl through empty drainpipes or even sewer pipes, and the power of flight enables them to get into a building through any number of openings.

The best way to prevent a roach infestation is to practice good house maintenance. Keeping the dwelling clean, especially in areas where food is stored, is almost 80% of the battle. Caulking cracks around pipes and filling possible entryways will also keep them from gaining access. If a solitary roach is seen when the house is being properly maintained, it should be eliminated and suspect areas should be sprayed immediately with a commercially available roach killer. The odds are good that a solitary roach found in a well-maintained home has gained access by hitching a ride on a parcel brought into the home. Killing them before they have time to reproduce can prevent an infestation.

Roaches do not really pose any threat to the structure of the house or to its occupants, although their droppings can contaminate food. Generally, people do not want their homes infested by bugs that cannot be easily controlled, if for none other than aesthetic reasons. This is where the adaptability of the roach becomes more aggravating. Commercial sprays have little effect on large roach populations because they are diluted so as not to harm humans or household pets. While spray will kill many roaches in the initial applications, the next generation of roaches spawned by the surviving adults will be more or less immune to the residual poison.

Another and more effective method of controlling roach populations is to place baited traps in strategic areas where roaches tend to breed and to gather food. By placing these poison traps around sinks, counters, stoves, and other areas where roaches may hide, their numbers can be controlled without the need for more drastic measures.

The most effective way to eliminate a roach problem is to hire a professional exterminator who will probably have to treat the home several times over the course of a year. Even professional strength sprays can have only a temporary effect on these stubborn bugs, but repeated applications usually achieve the desired results.

Home inspectors will recognize the presence of roaches without a great deal of difficulty. They will leave droppings in the areas where they hide and, by moving aside objects under which they may take shelter, one can discover carcasses of dead roaches. Live roaches are rarely seen during the daytime unless their hiding places are disturbed.

Fleas

Cats and dogs are commonly plagued by fleas, shown in Figure 9-8, which will spread to every section of the household if measures are not taken to prevent their entry and to eliminate them should they gain entry. Most of the time fleas can be eliminated with sprays available at most grocery stores. Flea collars and shampoos for pets will help to keep these irritating parasites out of the home. However, if some breeders (or unhatched eggs) remain, the next generation seems to become immune to the poisons and reproduction will increase. In many cases, a professional exterminator must be called in for a thorough exterminating job.

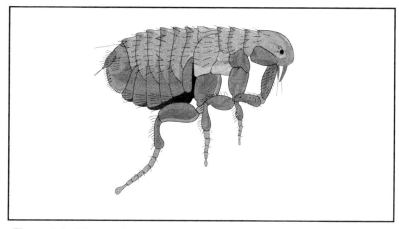

Figure 9-8: A home that has pets can easily become infested with fleas. They are hard to kill, and several treatments by experts may be necessary to get rid of fleas.

VERMIN

Keeping rats and mice out of the home is much like trying to keep out roaches. Good housekeeping will make food scarce and therefore make the home seem inhospitable to the rodents. Covering openings from the outside that are more than ¼" in diameter with metal, or stuffing them with steel wire will help keep them out.

Birds and bats can enter through holes in attic windows or other lofty areas that might discourage habitation by other animals. Keeping these pests out of the home is relatively easy. Since they most often enter through broken windows or through attic ventilation holes, good home maintenance will prevent them from moving in. Repairing broken windows immediately and putting screen over openings which may allow them to gain access is really all that is needed to prevent birds and bats from entering the home.

Squirrels will enter the home looking for shelter or for a place to store their winter food supply. Unfortunately, they are not extremely intelligent animals and, being unable to find the entrance through which they came, they will chew holes in wooden structures to form an exit. Red squirrels are usually more destructive than gray squirrels.

WOOD DECAY AND RADON GAS

Dry rot is caused by a fungus growing within the wooden structures of a building. In spite of its name, the fungus does best in wood that is wet most of the time, such as wooden framing members located near a leaking pipe or a basement wall through which water enters. Dry rot can also be found in attics where the roof is in poor shape and allows rain water to enter on a regular basis.

Wood that has been affected by dry rot can be probed easily with an ice pick or a screwdriver. There is little that can be done to save the wood that has already been affected if the fungi has done severe damage. Adding new wood is only a temporary solution as the fungus will spread into the new wood if conditions remain favorable. It is recommended that the wood be replaced and that the water supply allowing the fungus to receive moisture be repaired.

For most fungi to grow on wood, the wood must be almost constantly in contact with water. In fact, the wood must be almost completely saturated to allow the fungi spores to grow at all. Fungi survive by deriving nutrients from the substance on which they grow. They do not require much light and can grow in almost total darkness if there is enough moisture to sustain them.

Fungi damage wood by sending a "root" system throughout the wood and essentially digesting the fibers with which they come into contact. The roots secrete chemicals which dissolve the wood into a substance that can be absorbed by the body of the fungus. The fungal body may be in the form of a mushroom, tiny fruiting bodies, slime, or a scaled substance that sometimes resembles bird droppings.

In some cases there may be no visible sign of the presence of fungi. If the inspector suspects that fungi may have infiltrated the wood it is recommended that the wood be probed with an ice pick or a screwdriver. The wood will be spongy when wet and easily broken when dry. The affected wood will have an almost honeycombed appearance where the enzymes secreted by the fungus have dissolved the wood fibers.

Fungi can be difficult to control because they spread by way of spores, which are more like miniature dormant versions of fungi than seeds. These spores cannot be killed with chemicals once they have

Insects, Vermin and Decay

been formed within the bodies of the fungus, but they can be prevented from developing into the mature form of the fungus.

Radon Gas

Radon is a colorless, odorless, tasteless, radio active gas that comes from the natural breakdown of uranium. It can be found in most rocks and soils. Outdoors, it mixes with the air and is found in low concentrations that do not harm people. But indoors, it can accumulate and build up to levels that are dangerous.

High levels of radon in the home can increase the risk of developing lung cancer. At the present time, this is the only known adverse health effect. The Surgeon General's announcement in the 1980s drew attention to the dangers of radon by announcing that it is second only to smoking as a cause of lung cancer. Consequently, radon-testing kits became available so that homeowners could perform their own test, send off the sample, and get a report on the amount of radon in their home.

But how does radon get into the house? In general, the amount of radon in a particular home depends on the home's construction and the concentration of radon in the soil underneath it. Radon can enter a house through dirt floors, cracks in concrete foundations, floors and walls, floor drains, tiny cracks or pores in hollow-block walls, loose-fitting pipes, exhaust fans, sump pumps, and many other places, including the water supply.

The variation in radon levels has to do with the air-tightness of a house. The more energy efficient a home, the more likely that it will have higher radon levels. In the average house, there is one complete air exchange every six to seven hours. That is, about four times a day all the air from inside the house is exchanged with outside air. The tighter the house, the more likely it is that the air exchange will come from beneath the house, from the air over the soil that may contain high levels of radon gas.

To avoid high levels of radon gas, tightly seal off areas in which air from the ground can enter the house. Foundation walls should be checked for cracks, and any found should be tightly sealed. The seam between foundations walls and the concrete slab should be filled with a waterproof sealer. All holes or chases in the foundation walls, used for pipes and duct work, should be sealed.

Homes with crawl spaces can be helped by the installation of a continuous vapor barrier laid on the ground under the first floor, along with good crawl-space ventilation.

Increasing the number of air changes within the house can be accomplished by blowers and fans bringing in outside air and then exhausting it. However, in doing so, this make-up air will have to be preheated during cold weather. Obviously, this will increase the fuel bill for the home, but the additional expense is offset by the lesser risk of lung cancer.

Figure 9-9 shows the various areas where radon gas can enter the home. A radon testing kit placed at any of these areas should indicate the amount of radon entering the home.

INSPECTING THE HOME FOR PESTS AND ROT

Damage from termites, carpenter bees, carpenter ants, and wood rot is usually found in homes at least five years old, and more often, those 10-20 years old. Newer homes are almost always provided with termite shields and treated lumber. In newer homes, pests won't be able to penetrate the protected barriers as easily as in older homes.

When inspecting the foundation, look for flattened mud tunnels originating from the ground and following a path up the foundation wall. If such tunnels are detected, further investigation is necessary. Probe any suspicious wood members with an ice pick or screwdriver. Either of these tools will readily sink into a beam or joist if the inside is eaten away by any insect or wood decay. Also look for evidence of white wings atop the foundation wall and sawdust laying around in spots; both are indications of unwanted insects.

Another place to probe is at any water-stained or water-logged wood members. These are likely candidates for wood decay. If any evidence is found during your inspection, you should recommend to your client that the entire house be examined by an expert exterminator and obtain treatment. Remember, it is not the home inspector's job to correct defects; only to detect and report on them.

Damage by rodents is easily detected when gnawed holes are found in the building structure, especially around doors, windows, wooden vents, soffits, and similar locations. Also look for rodent droppings. Often, the type of rodent can be identified by these droppings or pellets.

Insects, Vermin and Decay

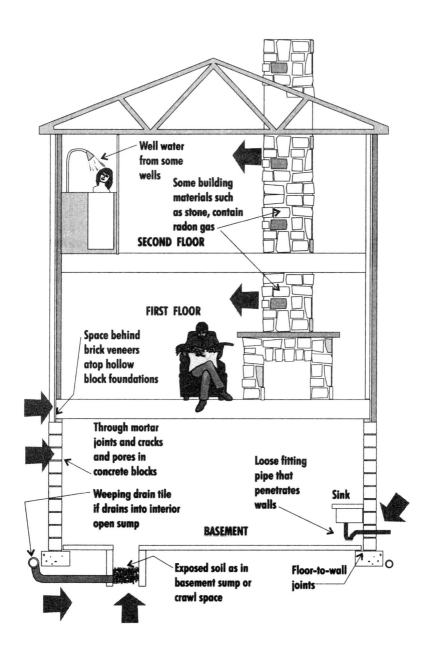

Figure 9-9: Areas where radon gas enters the home.

The home inspector can determine if there is a rodent problem by removing all pellets from a suspect area and then looking again the next day. If there are more pellets, there is a definite problem. Again, this is a job for a qualified exterminator.

Since radon gas is colorless, odorless, and tasteless, the only way to determine the amount of radon gas in a particular home is by using one of the many radon-testing kits that are available. In general, the testing device is placed in a suspicious area of the home, and left for approximately 30 days. The testing device is then packed up according to instructions accompanying the kit, and mailed to a testing lab. A report is usually made within 30 days.

The checklist in Figure 9-10 should facilitate your inspection for insects, vermin, and wood decay throughout the house. In almost every case, however, if suspicious areas are found, a qualified exterminator should be called in to verify your findings.

Insects, Vermin and Decay

	INSPECTION CHECKLIST — INSECTS, VERMIN, AND DECAY	Yes	No
1.	Are there any old books, clothing, or old wood on the ground near the house?		
2.	Are any flattened mud tunnels visible on foundation walls?		
3.	Did you find any silvery wings in the basement or crawl space area?		
4.	Did you find any sawdust laying around that could have been made by carpenter ants or bees?		
5.	Did you find any weak wooden members when probing with an ice pick or screwdriver?		
6.	Are there any gnawed holes in the soffits, eaves, around outside doors, or other places?		
7.	Did you find any rodent droppings during your inspection?		
8.	Are there any damp or wet wooden members in the house?		
9.	After probing any wet wooden members found, were they sound?		
10.	Has there been a recent termite check by a qualified exterminator?		
11.	Are wood joists, rafters, and other wooden members free of termites, carpenter ants, or wood rot?		
12.	Has the house been tested for radon gas?		
	Action that needs to be taken:		

Figure 9-10: Inspection checklist for insects, vermin, and decay.

Chapter 10
ELECTRICAL SYSTEMS

An overall understanding of the nature of electricity and of the system that produces it is the best foundation for understanding the inspection process of home electrical systems.

The essential elements of an electrical system capable of producing power to operate electrical appliances and lighting include generating stations, transformers, substations, transmission lines and distribution lines. The drawing in Figure 10-1 on the next page shows these elements and their relationships.

Electricity And Generation

Electricity is the flow of electrons, tiny atomic particles. These particles are found in all atoms. Atoms of some metals such as copper and aluminum have electrons that are easily pushed and guided into a stream. When a coil of metal wire is turned near a magnet, or vice versa, electricity will flow in the wire. This principle is made use of in generating plants; water or steam is used to turn turbines which rotate electromagnets that are surrounded by huge coils of wire. The push transmitted to the electrons by the turbine/magnet setup is measured in units called *volts*. The quantity of the flow of electricity is called *current* and it is measured in *amperes* or *"amps."*

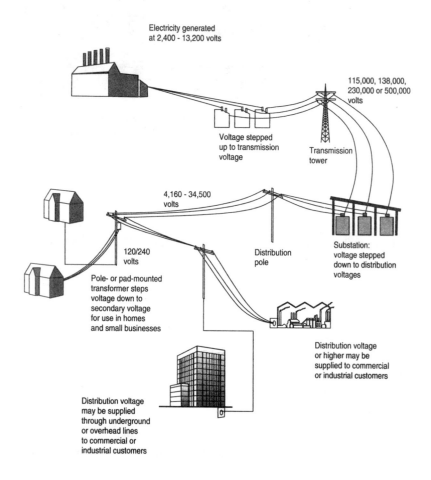

Figure 10-1: Sections of an electrical distribution system.

Electrical Systems

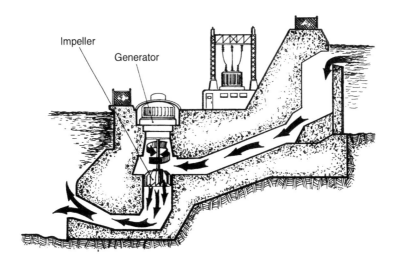

Figure 10-2: Hydro-electric generating plant.

Multiply volts by amps and the result is *watts* — the amount of work that electricity can do. In recent years however, the watt is more often called *volt-ampere*, since volts × amperes = watts. Electrical appliances and motors have certain watt requirements depending on the task they are expected to perform. When speaking of requirements for larger systems, the term *kilowatts*, (one kilowatt equals 1,000 watts) is used. This term is also used when speaking of power production or power needs. An electrical power plant produces kilowatts and power companies sell power in units called *kilowatthours*. For example, a small 100-watt fan motor operated for ten hours uses one kilowatt-hour of electricity (100 × 10 = 1000 watts or one kilowatt-hour).

Electricity is produced at the generating plant (Figure 10-2) at voltages varying from 2,400 volts to 13,200 volts. Transformers are also located at the generating plant to step up the voltage to hundreds of thousands of volts for transmission — a kind of wholesale block technique for economically moving large amounts of power from the generation point to key locations.

Electricity is transported from one part of the system to another by metal conductors, cables made up of many strands of wire. A continuous system of conductors through which electricity flows is called a circuit.

Transmission

The system for moving high voltage electricity is called the *transmission* system. Transmission lines are interconnected to form a network of lines. Should one line fail, another will take over the load. Such interconnections provide a reliable system for transporting power from generating plants to communities.

Most transmission lines installed by power companies utilize *three-phase current* — three separate streams of electricity, traveling on separate conductors. This is an efficient way to transport large quantities of electricity. At various points along the way, transformers step down the transmission voltage at facilities known as substations.

Substations can be small buildings or fenced in yards containing switches, transformers, and other electrical equipment and structures. Substations (Figure 10-3) are convenient places to monitor the system and adjust circuits. Devices called *regulators,* which maintain system voltage as the demand for electricity changes, are also installed in substations. Another device, which momentarily stores energy, is called a *capacitor,* and is sometimes installed in substations; this device reduces energy losses and improves voltage regulation. Within the substation, rigid tubular or rectangular bars, called *busbars* or *buses,* are used as conductors.

At the substation, the transmission voltage is stepped down to voltages below 69,000 volts which feed into the distribution system.

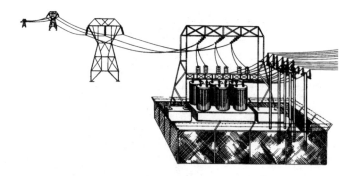

Figure 10-3: Typical substation.

Electrical Systems

The distribution system delivers electrical energy to the user's energy consuming equipment — such as lighting, motors, and of course, home appliances.

Conductors called *feeders,* radiating in all directions from the substation, carry the power from the substation to various distribution centers. At key locations in the distribution system, the voltage is stepped down by transformers to the level needed by the customer. Distribution conductors on the high voltage side of a transformer are called *primary conductors (primaries)*; those on the low voltage side are called *secondary conductors (secondaries).*

Transformers are smaller versions of substation regulators and capacitors are installed on poles throughout the distribution system.

Distribution lines carry either *three-phase* or *single-phase current.* Single-phase power is normally used for residential and small commercial occupancies, while three-phase power serves most of the other users.

Underground

Most power companies now utilize transmission systems that include both overhead and underground installations. In general, the terms and the devices are the same for both. In the case of the underground system, distribution transformers are installed at or below ground level. Those mounted on concrete slabs are called *padmounts,* (Figure 10-4) while those installed in underground vaults are called *submersibles.*

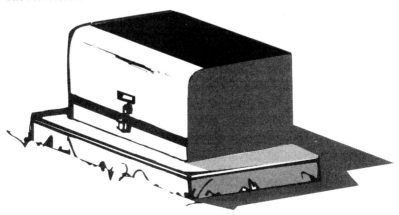

Figure 10-4: Padmount transformer.

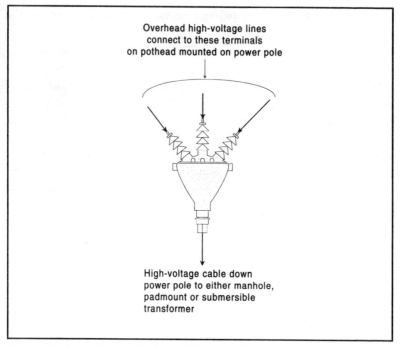

Figure 10-5: Typical pole-mounted pothead.

Buried conductors (cables) are insulated to protect them from soil chemicals and moisture. Many overhead conductors do not require such protective insulation.

When underground transmission or distribution cables terminate and connect with overhead conductors at buses or on the tops of poles, special devices called *potheads* or *cable terminators* are employed. These devices prevent moisture from entering the insulation of the cable and also serve to separate the conductors sufficiently to prevent *arcing* between them. The cable installation along the length of the pole is known as the *cable riser*. See Figure 10-5.

Secondary Systems

From a practical standpoint, those involved with home inspection need only be concerned with the power supply on the secondary (usage) side of the transformer, as this determines the characteristic of the power supply for use in the building or on the premises.

Two general arrangements of transformers and secondaries are in common use. The first arrangement is the sectional form, in which a unit of load, such as one city street or city block, is served by a fixed length of secondary conductors, with the transformer located in the middle. The second arrangement is the continuous form in which the secondary is installed in one long continuous run, with transformers spaced along it at the most suitable points. As the load grows or shifts, the transformers spaced along it can be moved or rearranged, if desired. In sectional arrangement, such a load can be cared for only by changing to a larger size of transformer or installing an additional unit in the same section.

One of the greatest advantages of the secondary bank is that the starting currents of motors, many of which are used in homes for washers, dryers, and other appliances, are divided among transformers, reducing voltage drop and also diminishing the resulting lamp flicker at the various outlets.

Power companies all over the United States and Canada are now trying to incorporate networks into their secondary power systems, especially in areas where a high degree of service reliability is necessary. Around cities and industrial applications, most secondary circuits are three phase, either 120/208 V or 480/208 V and wye-connected. Usually, two to four primary feeders are run into the area, and transformers are connected alternately to them. The feeders are interconnected in a grid, or network, so that if any feeder goes out of service the load is still carried by the remaining feeders.

The primary feeders supplying networks are run from substations at the usual primary voltage for the system, such as 4160, 4800, 6900, or 13,200 V. Higher voltages are practicable if the loads are large enough to warrant them.

Common Power Supplies

The most common power supply used for residential and small commercial applications is the 120/240 volt, single-phase service; it is used primarily for light and power, including single-phase motors up to about $7\frac{1}{2}$ horsepower (hp). A diagram of this service is shown in Figure 10-6 on the next page.

Four-wire delta-connected secondaries (Figure 10-7) and four-wire, wye-connected secondaries (Figure 10-8) are common around indus-

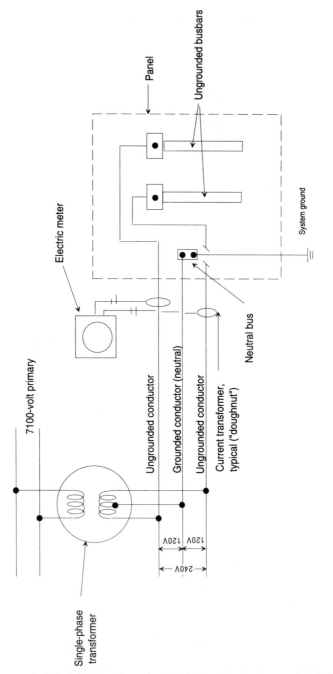

Figure 10-6: Single-phase, three-wire, 120/240-volt electric service.

Electrical Systems

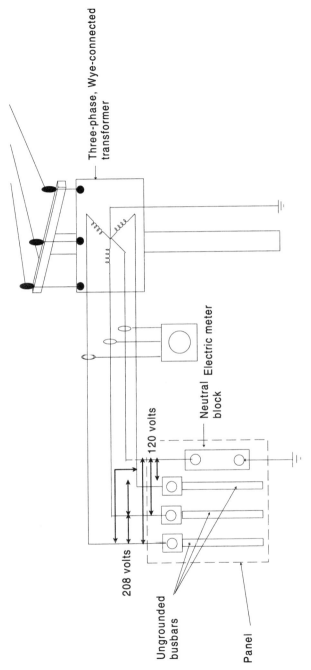

Figure 10-7: Three-phase, four-wire, wye-connected service.

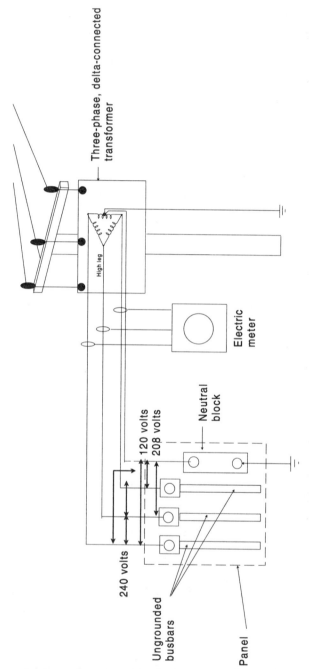

Figure 10-8: Four-wire, delta-connected service.

trial and large commercial applications, but the home inspector will have little opportunity to get involved with these systems.

Summary

The main purpose of this section is to give the home inspector an overall "feel" of today's electrical distribution systems and how electricity is generated and carried to the point where it will be utilized to operate electrical items in the home. The sections to follow will dig deeper into home electrical systems to give you a thorough understanding of all components and their characteristics as they relate to the home inspector. Such knowledge will enable you to perform your job in a more skillful way, and within the shortest amount of time.

It is not necessary for you to thoroughly understand electrical distribution at this time. If you have only a brief general understanding of how electricity is generated and transmitted, you have a good foundation for understanding the material presented in this chapter, and to becoming thoroughly knowledgeable in the field of inspecting home electrical systems.

To further support the knowledge you have already learned from this section, the next time you have the chance, observe the overhead electrical wiring in your area. Can you identify transformers mounted on the power poles? Can you distinguish capacitors from transformers? Look at the drawings presented in this section, and them compare them with the components on the distribution systems in your area. Before long, their knowledge will become second nature to you.

THE NATIONAL ELECTRICAL CODE®

Owing to the potential fire and explosion hazards caused by the improper handling and installation of electrical wiring, certain rules in the selection of materials, quality of workmanship, and precautions for safety must be followed. To standardize and simplify these rules and provide a reliable guide for electrical construction, the National Electrical Code (NEC) was developed. The NEC, originally prepared in 1897, is frequently revised to meet changing conditions, improved equipment and materials, and new fire hazards. It is a result of the best efforts of electrical engineers, manufacturers of electrical equipment,

Figure 10-9: The National Electrical Code has become the Bible of the electrical industry.

insurance underwriters, fire fighters, and other concerned experts throughout the country.

The *NEC* (Figure 10-9) is now published by the National Fire Protection Association (NFPA), Batterymarch Park, Quincy, Massachusetts 02269. It contains specific rules and regulations intended to help in the practical safeguarding of persons and property from hazards arising from the use of electricity.

Although the *NEC* itself states, "This Code is not intended as a design specification nor an instruction manual for untrained persons," it does provide a sound basis for the study of electrical design and installation procedures — under the proper guidance. The probable reason for the *NEC's* self-analysis is that the code also states, "This Code contains provisions considered necessary for safety. Compliance therewith and proper maintenance will result in an installation essen-

Electrical Systems

tially free from hazard, but not necessarily efficient, convenient, or adequate for good service or future expansion of electrical use."

The *NEC*, however, has become the Bible of the electrical construction industry, and anyone involved in home inspection, in *any* capacity, should obtain an up-to-date copy, keep it handy at all times, and refer to it frequently.

NEC TERMINOLOGY

There are two basic types of rules in the *NEC*: mandatory rules and advisory rules. Here is how to recognize the two types of rules and how they relate to all types of electrical systems.

- Mandatory rules—All mandatory rules have the word shall in them. The word "shall" means must. If a rule is mandatory, you must comply with it.

- Advisory rules—All advisory rules have the word should in them. The word "should" in this case means recommended but not necessarily required. If a rule is advisory, compliance is discretionary. If you want to comply with it, do so. But they are not mandatory.

Be alert to local amendments to the *NEC*. Local ordinances may amend the language of the *NEC*, changing it from should to shall. This means that you must do in that county or city what may only be recommended in some other area. The office that issues building permits will either sell you a copy of the code that's enforced in that area or tell you where the code is sold. In rare instances, the electrical inspector having jurisdiction over the area, may issue these regulations verbally.

There are a few other "landmarks" that you will encounter while looking through the *NEC*. These are summarized in Figure 10-10 on the next page, and a brief explanation of each follows:

Explanatory Material: Explanatory material in the form of Fine Print Notes is designated (FPN). Where these appear, the FPNs normally apply to the *NEC* Section or paragraph immediately preceding the FPN.

```
┌─────────────────────────────────────────┐
│     Mandatory rules are characterized by │
│          the use of the word:            │
│                 SHALL                    │
│                                          │
│     A recommendation or that which is    │
│         advised but not required is      │
│    characterized by the use of the word: │
│                SHOULD                    │
│                                          │
│       Explanatory material in the form of│
│         Fine Print Notes is designated:  │
│                 (FPN)                    │
│    │ A change bar in the margins         │
│    │ indicates that a change in the      │
│    │ NEC has been made since the         │
│      last edition.                       │
│      A bullet indictates that something  │
│    ● has been deleted from the last      │
│      edition of the NEC.                 │
└─────────────────────────────────────────┘
```

Figure 10-10: Summary of NEC terminology.

Change Bar: A change bar in the margins indicates that a change in the NEC has been made since the last edition. When becoming familiar with each new edition of the *NEC*, always review these changes. There are also several illustrated publications on the market that point out changes in the *NEC* with detailed explanations of each. Such publications make excellent reference material.

Bullets: A filled-in circle called a "bullet" indicates that something has been deleted from the last edition of the *NEC*. Although not absolutely necessary, many electricians like to compare the previous *NEC* edition to the most recent one when these bullets are encountered, just to see what has been omitted from the latest edition. The most probable reasons for the deletions are errors in the previous edition, or obsolete items.

Extracted Text: Material identified by the superscript letter "x" includes text extracted from other NFPA documents as identified in Appendix A of the *NEC*.

As you open the *NEC* book, you will notice several different types of text used. Here is an explanation of each.

Electrical Systems

Grounding Electrode Conductor for AC Systems			
Size of Largest Service-Entrance Conductor or Equivalent Area for Parallel Conductors		Size of Grounding Electrode Conductor	
Copper	Aluminum or Copper-Clad Aluminum	Copper	Aluminum or Copper-Clad Aluminum
2 or smaller	1/0 or smaller	8	6
1 or 1/0	2/0 or 3/0	6	4
2/0 or 3/0	4/0 or 250 kcmil	4	2
Over 3/0 thru 350 kcmil	Over 250 kcmil thru 500 kcmil	2	1/0

Figure 10-11: Typical *NEC* table.

1. *Black Letters:* Basic definitions and explanations of the *NEC*.
2. *Bold Black Letters:* Headings for each *NEC* application.
3. *Exceptions:* These explain the situations when a specific rule does not apply. Exceptions are written in italics under the Section or paragraph to which they apply.
4. *Tables:* Tables are often included when there is more than one possible application of a requirement. See Figure 10-11.
5. *Diagrams:* A few diagrams are scattered throughout the *NEC* to illustrate certain *NEC* applications. See Figure 10-12 on the next page.

LEARNING THE NEC LAYOUT

The *NEC* is divided into the Introduction (Article 90) and nine chapters. Chapters 1, 2, 3, and 4 apply generally; Chapters 5, 6, and 7 apply to special occupancies, special equipment, or other special conditions. These latter chapters supplement or modify the general rules. Chapters 1 through 4 apply except as amended by Chapters 5, 6, and 7 for the particular conditions.

While looking through these NEC chapters, if you should encounter a word or term that is unfamiliar, look in Chapter 1, Article 100 —

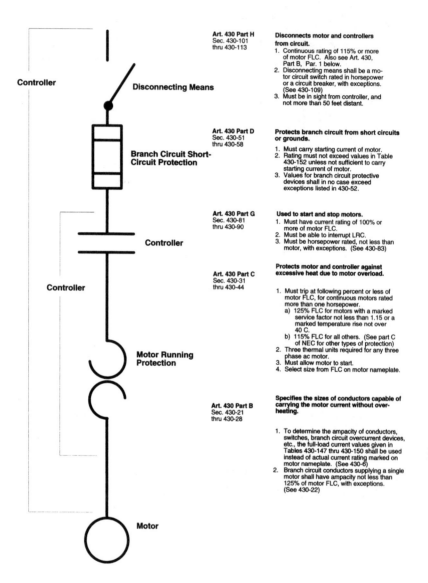

Figure 10-12: The NEC contains a few diagrams to better explain certain wiring requirements.

Electrical Systems

Definitions. Chances are, the term will be found here. If not, look in the Index for the word and the *NEC* page number. Many terms are included in Article 100, but others are scattered throughout the book.

For definitions of terms not found in the *NEC*, obtain a copy of *Illustrated Dictionary for Electrical Workers*, available from Delmar Publishers, Inc., Albany, NY.

Chapter 8 of the *NEC* covers communications systems and is independent of the other chapters except where they are specifically referenced therein.

Chapter 9 consists of tables and examples.

There is also the *NEC* Contents at the beginning of the book and a comprehensive Index at the back of the book. You will find frequent use for both of these helpful "tools" when searching for various installation requirements.

Each chapter is divided into one or more Articles. For example Chapter 1 contains Articles 100 and 110. These Articles are subdivided into Sections. For example, Article 110 of Chapter 1 begins with Section 110-2, Approval. A bullet in the margin indicates that Section 110-1 has been deleted from the last *NEC* edition. Some sections may contain only one sentence or a paragraph, while others may be further subdivided into lettered or numbered paragraphs such as (a), (1), (2), and so on.

Begin your study of the *NEC* with Articles 90, 100 and 110. These three articles have the basic information that will make the rest of the *NEC* easier to understand. Article 100 defines terms you will need to understand the code. Article 110 gives the general requirements for electrical installations. Read these three articles over several times until you are thoroughly familiar with all the information they contain. It's time well spent. For example, Article 90 contains the following sections:

- Purpose (90-1)

- Scope (90-2)

- Code Arrangement (90-3)

- Enforcement (90-4)

- Mandatory Rules and Explanatory Material (90-5)

- Formal Interpretations (90-6)

- Examination of Equipment for Safety (90-7)

- Wiring Planning (90-8)

Once you are familiar with Articles 90, 100, and 110 you can move on to the rest of the *NEC*. There are several key sections you will use often in servicing electrical systems. Let's discuss each of these important sections.

Wiring Design and Protection: Chapter 2 of the *NEC* discusses wiring design and protection, the information electrical technicians need most often. It covers the use and identification of grounded conductors, branch circuits, feeders, calculations, services, overcurrent protection and grounding. This is essential information for any type of electrical system, regardless of the type.

Chapter 2 is also a "how-to" chapter. It explains how to provide proper spacing for conductor supports, how to provide temporary wiring and how to size the proper grounding conductor or electrode. If you run into a problem related to the design/installation of a conventional electrical system, you can probably find a solution for it in this chapter.

Wiring Methods and Materials: Chapter 3 has the rules on wiring methods and materials. The materials and procedures to use on a particular system depend on the type of building construction, the type of occupancy, the location of the wiring in the building, the type of atmosphere in the building or in the area surrounding the building, mechanical factors and the relative costs of different wiring methods. See Figure 10-13.

The provisions of this article apply to all wiring installations except remote control switching (Article 725), low-energy power circuits (Article 725), signal systems (Article 725), communication systems and conductors (Article 800) when these items form an integral part of equipment such as motors and motor controllers.

Electrical Systems

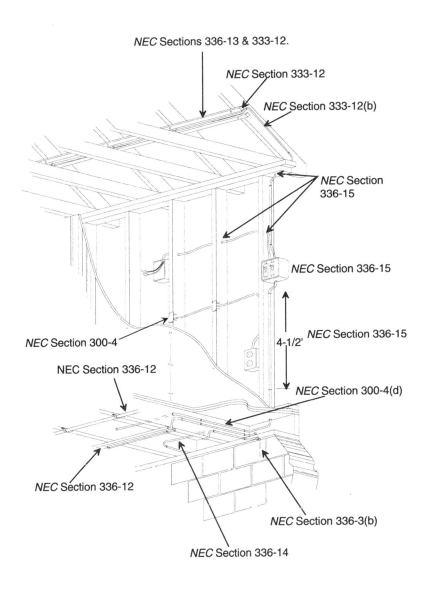

Figure 10-13: *NEC* **regulations governing wiring with Type NM (Romex) cable — the most popular wiring method for homes.**

There are four basic wiring methods used in most modern electrical systems. Nearly all wiring methods are a variation of one or more of these four basic methods:

- Sheathed cables of two or more conductors, such as NM cable and BX armored cable (Articles 330 through 339)

- Raceway wiring systems, such as rigid and EMT conduit (Articles 342 to 358)

- Busways (Article 364)

- Cabletray (Article 318)

Article 310 in Chapter 3 gives a complete description of all types of electrical conductors. Electrical conductors come in a wide range of sizes and forms. Be sure to check the working drawings and specifications to see what sizes and types of conductors are required for a specific job. If conductor type and size are not specified, choose the most appropriate type and size meeting standard *NEC* requirements.

Articles 318 through 384 give rules for raceways, boxes, cabinets and raceway fittings. Outlet boxes vary in size and shape, depending on their use, the size of the raceway, the number of conductors entering the box, the type of building construction and atmospheric conditions of the areas. Chapter 3 should answer most questions on the selection and use of these items.

The *NEC* does not describe in detail all types and sizes of outlet boxes. But manufacturers of outlet boxes have excellent catalogs showing all of their products. Collect these catalogs. They are essential to your work.

Article 380 covers the switches, push buttons, pilot lamps, receptacles and convenience outlets you will use to control electrical circuits or to connect portable equipment to electric circuits. Again, get the manufacturers' catalogs on these items. They will provide you with detailed descriptions of each of the wiring devices.

Article 384 covers switchboards and panelboards, including their location, installation methods, clearances, grounding and overcurrent protection.

Equipment For General Use

Chapter 4 of the *NEC* begins with the use and installation of flexible cords and cables, including the trade name, type letter, wire size, number of conductors, conductor insulation, outer covering and use of each. The chapter also includes fixture wires, again giving the trade name, type letter and other important details.

Article 410 on lighting fixtures is especially important. It gives installation procedures for fixtures in specific locations. For example, it covers fixtures near combustible material and fixtures in closets. The *NEC* does not describe how many fixtures will be needed in a given area to provide a certain amount of illumination.

Article 430 covers electric motors, including mounting the motor and making electrical connections to it. Motor controls and overload protection are also covered.

Articles 440 through 460 cover air conditioning and heating equipment, transformers and capacitors.

Article 480 gives most requirements related to battery-operated electrical systems. Storage batteries are seldom thought of as part of a conventional electrical system, but they often provide standby emergency lighting service. They may also supply power to security systems that are separate from the main ac electrical system.

Special Occupancies

Chapter 5 of the *NEC* covers special occupancy areas. These are areas where the sparks generated by electrical equipment may cause an explosion or fire. The hazard may be due to the atmosphere of the area or just the presence of a volatile material in the area. Commercial garages, aircraft hangers and service stations are typical special occupancy locations.

Articles 500 through 501 cover the different types of special occupancy atmospheres where an explosion is possible. The atmospheric groups were established to make it easy to test and approve equipment for various types of uses.

Articles 501-4, 502-4 and 503-3 cover the installation of explosion-proof wiring. An explosion-proof system is designed to prevent the ignition of a surrounding explosive atmosphere when arcing occurs within the electrical system.

There are three main classes of special occupancy locations:

- *Class I (Article 501):* Areas containing flammable gases or vapors in the air. Class I areas include paint spray booths, dyeing plants where hazardous liquids are used and gas generator rooms.

- *Class II (Article 502):* Areas where combustible dust is present, such as grain-handling and storage plants, dust and stock collector areas and sugar-pulverizing plants. These are areas where, under normal operating conditions, there may be enough combustible dust in the air to produce explosive or ignitable mixtures.

- *Class III (Article 503):* Areas that are hazardous because of the presence of easily ignitable fibers or flyings in the air, although not in large enough quantity to produce ignitable mixtures. Class III locations include cotton mills, rayon mills and clothing manufacturing plants.

Articles 511 and 514 regulate garages and similar locations where volatile or flammable liquids are used. While these areas are not always considered critically hazardous locations, there may be enough danger to require special precautions in the electrical installation. In these areas, the *NEC* requires that volatile gases be confined to an area not more than 4 feet above the floor. So in most cases, conventional raceway systems are permitted above this level. If the area is judged critically hazardous, explosionproof wiring (including seal-offs) may be required.

Article 520 regulates theaters and similar occupancies where fire and panic can cause hazards to life and property. Drive-in theaters do not present the same hazards as enclosed auditoriums. But the projection rooms and adjacent areas must be properly ventilated and wired for the protection of operating personnel and others using the area.

Chapter 5 also covers residential storage garages, aircraft hangars, service stations, bulk storage plants, health care facilities, mobile homes and parks, and recreation vehicles and parks.

When installing electrical systems in Class I, Division 1 locations, explosionproof fittings are required and most electrical wiring must be enclosed in rigid steel conduit (pipe).

Special Equipment

Home inspectors will seldom need to refer to the Articles in Chapter 6 of the *NEC*, but the items in Chapter 6 are frequently encountered by commercial and industrial electrical workers.

Article 600 covers electric signs and outline lighting. Article 610 applies to cranes and hoists. Article 620 covers the majority of the electrical work involved in the installation and operation of elevators, dumbwaiters, escalators and moving walks. The manufacturer is responsible for most of this work. The electrician usually just furnishes a feeder terminating in a disconnect means in the bottom of the elevator shaft. The electrician may also be responsible for a lighting circuit to a junction box midway in the elevator shaft for connecting the elevator cage lighting cable and exhaust fans. Articles in Chapter 6 of the *NEC* give most of the requirements for these installations.

Article 630 regulates electric welding equipment. It is normally treated as a piece of industrial power equipment requiring a special power outlet. But there are special conditions that apply to the circuits supplying welding equipment. These are outlined in detail in Chapter 6 of the *NEC*.

Article 640 covers wiring for sound-recording and similar equipment. This type of equipment normally requires low-voltage wiring. Special outlet boxes or cabinets are usually provided with the equipment. But some items may be mounted in or on standard outlet boxes. Some sound-recording electrical systems require direct current, supplied from rectifying equipment, batteries or motor generators. Low-voltage alternating current comes from relatively small transformers connected on the primary side to a 120-volt circuit within the building.

Other items covered in Chapter 6 of the *NEC* include: X-ray equipment (Article 660), induction and dielectric heat-generating equipment (Article 665) and machine tools (Article 670).

If you ever have work that involves Chapter 6, study the chapter before work begins. That can save a lot of installation time. Here is another way to cut down on labor hours and prevent installation errors. Get a set of rough-in drawings of the equipment being installed. It is easy to install the wiring outlet box or to install the right box in the wrong place. Having a set of rough-in drawings can prevent those simple but costly errors.

Special Conditions

In most commercial buildings, the *NEC* and local ordinances require a means of lighting public rooms, halls, stairways and entrances. There must be enough light to allow the occupants to exit from the building if the general building lighting is interrupted. Exit doors must be clearly indicated by illuminated exit signs.

Chapter 7 of the *NEC* covers the installation of emergency lighting systems. These circuits should be arranged so that they can automatically transfer to an alternate source of current, usually storage batteries or gasoline-driven generators. As an alternative in some types of occupancies, you can connect them to the supply side of the main service so disconnecting the main service switch would not disconnect the emergency circuits. See Article 700. *NEC* Chapter 7 also covers a variety of other equipment, systems and conditions that are not easily categorized elsewhere in the *NEC*.

Chapter 8 is a special category for wiring associated with electronic communications systems including telephone and telegraph, radio and TV, fire and burglar alarms, and community antenna systems.

USING THE NEC

Once you become familiar with the *NEC* through repeated usage, you will generally know where to look for a particular topic. While this chapter provides you with an initial familiarization of the *NEC* layout, much additional usage experience will be needed for you to feel comfortable with the *NEC's* content. Here's how to locate information on a specific subject.

Step 1. Look through the Contents. You may spot the topic in a heading or subheading. If not, look for a broader, more general subject heading under which the specific topic may appear. Also look for related or similar topics. The Contents will refer you to a specific page number.

Step 2. If you do not find what you're looking for in the Contents, go to the Index at the back of the book. This alphabetic listing is finely divided into different topics. You should locate the subject here. The Index, how-

Electrical Systems 211

ever, will refer to you either an Article or Section number (not a page number) where the topic is listed.

Step 3. If you cannot find the required subject in the Index, try to think of alternate names. For example, instead of wire, look under conductors; instead of outlet box, look under boxes, outlet, and so on.

The *NEC* is not an easy book to read and understand at first. In fact, seasoned electrical workers and technicians may find it confusing. Basically, it is a reference book written in a legal, contract-type language and its content does assume prior knowledge of most subjects listed. Consequently, you will sometimes find the *NEC* frustrating to use because terms aren't always defined, or some unknown prerequisite knowledge is required. To minimize this problem, it is recommended that you obtain one of the several *NEC* supplemental guides that are designed to explain and supplement the *NEC*. One of the best is *The National Electrical Code Handbook*, available from the NFPA, Batterymarch Park, Quincy, MA 02269 or from your local book store. Another excellent *NEC* cross-reference is *Illustrated Code-X, Encyclopedic Cross-Reference to the NEC,* available from John E. Traister Associates, P.O. Box 300, Bentonville, Virginia 22610 for $45.00 post paid.

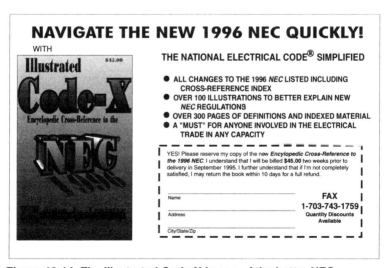

Figure 10-14: The Illustrated Code-X is one of the better NEC supplemental references.

Practical Application

Let's assume that a 120-volt outlet box is to be installed to provide the power supply for a surveillance camera in a commercial office. The owner wants the outlet box surface-mounted and located behind a curtain of their sliding glass patio doors. To determine if this is an *NEC* violation or not, follow these steps:

Step 1. Turn to the Contents of the *NEC* book, which begins on page 70-V.

Step 2. Find the chapter that would contain information about the general application you are working on. For this example, Chapter 4—Equipment for General Use should cover track lighting.

Step 3. Now look for the article that fits the specific category you are working on. In this case, Article 410 covers lighting fixtures, lampholders, lamps, and receptacles.

Step 4. Next locate the NEC Section within the *NEC* Article 410 that deals with the specific application. For this example, refer to Part R—Lighting Track.

Step 5. Turn to the page listed. The 1993 *NEC* gives page 350.

Step 6. Read *NEC* Section 410-100, Definition to become familiar with track lighting. Continue down the page with *NEC* Section 410-101 and read the information contained therein. Note that paragraph (c) under *NEC* Section 410-101 states the following:

(c) Locations Not Permitted. Lighting track shall not be installed (1) where subject to physical damage; (2) in wet or damp locations; (3) where subject to corrosive vapors; (4) in storage battery rooms; (5) in hazardous (classified) locations; (6) where concealed; (7) where extended through walls or partitions; (8) less than 5 feet above the finished floor except where protected from physical damage or track operating at less than 30 volts RMS open-circuit voltage.

Step 7. Read *NEC* Section 410-101, paragraph (c) carefully. Do you see any conditions that would violate any *NEC* requirements if the track lighting is installed in the area specified? In checking these items, you will probably

Electrical Systems 213

Step 8. note condition (6), "where concealed." Since the track lighting is to be installed behind a curtain, this sounds like an *NEC* violation. But let's check further.
Step 8. Let's get an interpretation of the *NEC's* definition of "concealed." Therefore, turn to Article 100 — definitions and find the main term "concealed." It reads as follows:

Concealed: Rendered inaccessible by the structure or finish of the building....

Step 9. After reading the *NEC's* definition of "concealed," although the track lighting may be out of sight (if the curtain is drawn), it will still be readily accessible for maintenance. Consequently, the track lighting is really not concealed according to the *NEC* definition.

When using the *NEC* to determine correct electrical-installation requirements, please keep in mind that you will nearly always have to refer to more than one Section. Sometimes the *NEC* itself refers the reader to other Articles and Sections. In some cases, the user will have to be familiar enough with the *NEC* to know what other *NEC* Sections pertain to the installation at hand. It's a confusing situation to say the least, but time and experience in using the *NEC* frequently will make using it much easier.

Now let's take another example to further acquaint you with navigating the *NEC*.

Suppose you are installing Type SE (service-entrance) cable on the side of a home. You know that this cable must be secured, but you aren't sure of the spacing between cable clamps. To find out this information, use the following procedure:

Step 1: Look in the *NEC* Table of Contents and follow down the list until you find an appropriate category.
Step 2: Article 230 under Chapter 3 will probably catch your eye first, so turn to the page where Article 230 begins in the *NEC*.

Step 3: Glance down the section numbers, 230-1, Scope, 230-2, Number of Services, etc. until you come to Section 230-51, Mounting Supports. Upon reading this section, you will find in paragraph (a) — Service -Entrance Cables — that "Service-entrance cable shall be supported by straps or other approved means within 12 inches (305 mm) of every service head, gooseneck, or connection to a raceway or enclosure and at intervals not exceeding 30 inches (762 mm)."

After reading this section, you will know that a cable strap is required within 12 inches of the service head and within 12 inches of the meter base. Furthermore, the cable must be secured in between these two termination points at intervals not exceeding 30 inches.

DEFINITIONS

Many definitions of terms dealing with the *NEC* may be found in *NEC* Article 100. However, other definitions are scattered throughout the *NEC* under their appropriate category. For example the term lighting track, as discussed previously, is not listed in Article 100. The term is listed under *NEC* Section 410-100 and reads as follows:

"Lighting track is a manufactured assembly designed to support and energize lighting fixtures that are capable of being readily repositioned on the track. Its length may be altered by the addition or subtraction of sections of track."

Regardless of where the definition may be located — in Article 100 or under the appropriate *NEC* Section elsewhere in the book — the best way to learn and remember these definitions is to form a mental picture of each item or device as you read the definition. For example, turn to page 70-5 of the 1993 *NEC* and under Article 100 — Definitions, scan down the page until you come to the term "Attachment Plug (Plug Cap) (Cap)." After reading the definition, you will probably have already formed a mental picture of attachment plugs. See Figure 10-15 on the next page for some of the more common attachment plugs.

Electrical Systems

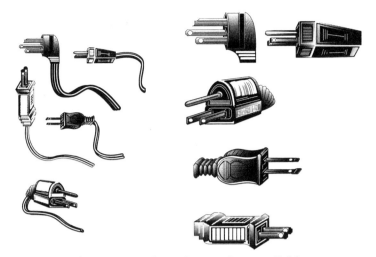

Figure 10-15: Some types of attachment plugs available.

Once again, scan through the definitions until the term "Appliance" is found. Read the definition and then try to form a mental picture of what appliances look like. Some of the more common appliances appear in Figure 10-16 on the next page. They should be familiar to everyone.

Each and every term listed in the *NEC* should be understood. Know what the item looks like and how it is used on the job. If a term is unfamiliar, try other reference books such as manufacturers' catalogs for an illustration of the item. Then research the item further to determine its purpose in electrical systems. Once you are familiar with all the common terms and definitions found in the *NEC*, navigating through the *NEC* (and understanding what you read) will be much easier.

TESTING LABORATORIES

There are many definitions included in Article 100. You should become familiar with the definitions. Since a copy of the latest *NEC* is compulsory for any type of electrical-wiring installation and inspection, there is no need to duplicate them here. However, here are two definitions that you should become especially familiar with:

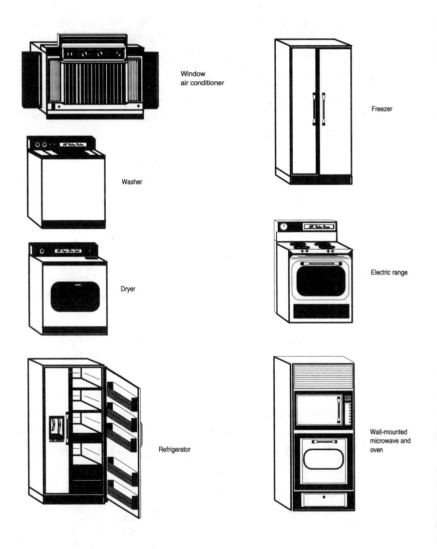

Figure 10-16: Common household appliances.

Electrical Systems

- Labeled - Equipment or materials to which has been attached a label, symbol or other identifying mark of an organization acceptable to the authority having jurisdiction and concerned with product evaluation, that maintains periodic inspection of production of labeled equipment or materials, and by whose labeling the manufacturer indicates compliance with appropriate standards or performance in a specified manner.

- Listed - Equipment or materials included in a list published by an organization acceptable to the authority having jurisdiction and concerned with product evaluation, that maintains periodic inspection of production of listed equipment or materials, and whose listing states either that the equipment or material meets appropriate designated standards or has been tested and found suitable for use in a specified manner. Besides installation rules, you will also have to be concerned with the type and quality of materials that are used in electrical wiring systems. Nationally recognized testing laboratories (Underwriters' Laboratories, Inc. is one) are product safety certification laboratories. They establish and operate product safety certification programs to make sure that items produced under the service are safeguarded against reasonable foreseeable risks. Some of these organizations maintain a worldwide network of field representatives who make unannounced visits to manufacturing facilities to countercheck products bearing their "seal of approval." See Figure 10-17.

Figure 10-17: UL label.

However, proper selection, overall functional performance and reliability of a product are factors that are not within the basic scope of UL activities.

To fully understand the *NEC*, it is important to understand the organizations which govern it.

NRTL (Nationally Recognized Testing Laboratory)

Nationally Recognized Testing Laboratories are product safety certification laboratories. They establish and operate product safety certification programs to make sure that items produced under the service are safeguarded against reasonable foreseeable risks. NRTL maintains a worldwide network of field representatives who make unannounced visits to factories to countercheck products bearing the safety mark.

NEMA (National Electrical Manufacturers Association)

The National Electrical Manufacturers Association was founded in 1926. It is made up of companies that manufacture equipment used for generation, transmission, distribution, control, and utilization of electric power. The objectives of NEMA are to maintain and improve the quality and reliability of products; to ensure safety standards in the manufacture and use of products; to develop product standards covering such matters as naming, ratings, performance, testing, and dimensions. NEMA participates in developing the *NEC* and the National Electrical Safety Code and advocates their acceptance by state and local authorities.

NFPA (National Fire Protection Association)

The NFPA was founded in 1896. Its membership is drawn from the fire service, business and industry, health care, educational and other institutions, and individuals in the fields of insurance, government, architecture, and engineering. The duties of the NFPA include:

- Developing, publishing, and distributing standards prepared by approximately 175 technical committees. These standards are intended to minimize the possibility and effects of fire and explosion.

Electrical Systems

- Conducting fire safety education programs for the general public.

- Providing information on fire protection, prevention, and suppression.

- Compiling annual statistics on causes and occupancies of fires, large-loss fires (over 1 million dollars), fire deaths, and firefighter casualties.

- Providing field service by specialists on electricity, flammable liquids and gases, and marine fire problems.

- Conducting research projects that apply statistical methods and operations research to develop computer modes and data management systems.

The Role Of Testing Laboratories

Testing laboratories are an integral part of the development of the code. The NFPA, NEMA, and NRTL all provide testing laboratories to conduct research into electrical equipment and its safety. These laboratories perform extensive testing of new products to make sure they are built to code standards for electrical and fire safety. These organizations receive statistics and reports from agencies all over the United States concerning electrical shocks and fires and their causes. Upon seeing trends developing concerning association of certain equipment and dangerous situations or circumstances, this equipment will be specifically targeted for research.

ELECTRIC CIRCUITS

The three basic electrical units in any electrical circuit are the ampere, volt, and ohm. The pressure in an electrical circuit is measured in volts, and may be compared to pressure in a water pipe except that water pressure is measured in pounds per square inch. The rate of current in an electrical circuit is measured in amperes instead of gallons per minute in a water pipe. Resistance in an electrical circuit, measured in ohms, is similar to friction in a water pipe.

Another important unit of electrical measurement is the *watt*, currently referred to in the *NEC* as *volt-ampere*, since volts times amperes = watts. Again, using a water supply system as an example, if you turn on a water faucet with a rating of 3 gallons per minute and leave it on for 5 minutes, you will have used (3 x 5=) 15 gallons of water. By the same token if you use a 120-volt appliance with a current rating of 10 amperes, the power consumed will be (120 x 10=) 1200 volt-amperes (sometimes called *watts*). Since current is charged by the kilowatt-hour, if this appliance was left on for one hour, you would pay for 1200 watt-hours of electricity. Since *kilo* means 1000, and you used 1200 watts, you would be charged for 1.2 kilowatt-hours. If the rate is, say, 8 cents per kilowatt-hour, you would be charged .0960 cents, or just about a dime for using a 1200-watt appliance for one hour.

The units just described would have no relationship with each other unless they were defined. A very important basic relationship among these units has been established and is called *Ohm's law*. This law states that current flow is proportional to voltage but is inversely proportional to resistance. In other words, the law states that the current, in amperes, increases and decreases directly with an increase or decrease of the pressure difference in volts. It further states that when the resistance is doubled, only half as much current will flow (when the voltage remains the same), and when half the resistance is present, twice as much current will flow.

The basic ways of stating Ohm's law when I = amperes, R = resistance in ohms, and E = volts are as follows:

- $E = IR$, or the voltage is equal to the current multiplied by the resistance.

- $I = E/R$, or the current equals the voltage divided by the resistance.

- $R = E/I$, or the resistance equals the voltage divided by the current in amperes.

The use of these equations enables us to calculate the third quantity if any two are already known.

Electrical Systems

The watt or volt-ampere may also be incorporated into Ohm's law for further calculations. When W = watts, current may be found by the following equations:

- $I = W/E$, or the current equals the wattage divided by the voltage.

- $I = \sqrt{W/R}$, or the current equals the square root of the wattage divided by the resistance.

Voltage may be found by using the following equations:

- $E = W/I$, or the voltage equals the wattage divided by the current.

- $E = \sqrt{W} \times R$, or the voltage equals the square root of the wattage times the resistance.

Resistance may be found by the following equations:

- $R = E^2/W$, or the resistance equals the voltage squared divided by the wattage.

- $R = W/I^2$, or the resistance equals the wattage divided by the current squared.

The power in watts (volt-amperes) of a circuit may be found by the following equations:

- $W = E^2/R$, or the wattage equals the voltage squared divided by the resistance.

- $W = I^2 \times R$, or the wattage equals the current squared times the resistance.

- $W = E \times I$, or the wattage equals the voltage times the current.

INSPECTING THE ELECTRICAL SYSTEM

The information in this section are suggestions for inspecting the home's electrical system. In many cases, these suggestions go beyond the normal scope of the home inspector's duties, as required by most lending institutions at the present time. However, the situation is likely to change in the very near future; that is, as the field of home inspection becomes more competitive, home inspectors will increase the breadth of their inspection as they gain more knowledge of building construction and the related systems within the home. Consequently, the information contained herein not only prepares you for the present, but also for the future.

Most electrical installations are governed by the National Electrical Code, but since the *NEC* is revised (with many changes) approximately every three years, homes built prior to this change may not be wired to current standards. So when you evaluate any home, keep this in mind; otherwise, you might report an unsuitable condition of the house wiring that is, in fact, okay. In other words, in existing homes that have once been approved by the local electrical inspector, the *NEC* does not require these homes to be upgraded each time a new edition of the Code is printed. For example, numerous homes were wired with the knob and tube method up until about 1940. The *NEC* now allows this wiring method only in farm and industrial buildings. However, it does not require that existing installations be upgraded. Furthermore, existing installations may be extended in the case of new additions to the home and also maintained and retained in service. Therefore, when you inspect older homes that have been renovated or new additions added, you may find more than one wiring method. As long as the methods were approved at the time of their installation, these are fully acceptable as far as the home inspector is concerned.

Electric Services

In general, there are two basic types of electric services: overhead and underground, or a combination of the two. When you approach the property to be inspected, observe any nearby power lines. If the home in question has an overhead service-entrance, the wiring from a nearby power pole to the house should be very conspicuous.

Electrical Systems

If no power lines are visible, chances are the home is utilizing an underground service-entrance. To be sure however, locate the power company's meter on the house. If there is a conduit (pipe) running out of the bottom of the meter into the ground, and no other visible connections to the meter, it is certain to be an underground service. An overhead service should be plainly visible.

Overhead service conductors must be readily accessible. They must have a clearance of not less than 3 feet from windows, doors, porches, fire escapes, or similar locations. Furthermore, when service conductors pass over rooftops, they must have a clearance of not less than 8 feet from the highest points of roofs. There are, however, some exceptions.

The *NEC* also specifies the distance that service conductors must clear the ground. These distances will vary with the surrounding conditions.

In general, service conductors must be at least 10 feet above the ground or other accessible surfaces at all times. If the service passes over residential property and driveways not subject to truck traffic, the conductors must be at least 12 feet above the ground. If subject to truck traffic, the distance must be increased to 18 feet.

Your checklist thus far should indicate the following:

- Overhead or underground service.

- Location of meter.

- Number of conductors entering service head.

- If overhead service, approximate height above ground.

You should now examine the main panelboard which is located on the inside of the home. Always carry a flashlight during this inspection because many times the panelboard (main load center) is located in a dimly lighted area, like in the utility room in a basement.

Once located, read any nameplates visible on the outside. You should be able to determine the ampere size of the main switch or panelboard from this. If none is located on the outside of the housing, over the housing door; the nameplate should be located on the inside of the door. In homes built after the 1960s, the panel should be

120/240-volts, single-phase and rated at either a 100- or 200-ampere capacity. Homes built before the 1960s will probably be rated for 60 amperes. In rare cases, you may still find a 120-volt, single-phase, 30-ampere switch which were common during the early part of this century before electrical appliances came into play. If so, this should be upgraded to 100 amperes.

While you have the panelboard door open, determine if the overcurrent protection consists of fuses or circuit breakers. Also look for any brown or black smudges around the overcurrent devices; this is a sure indication of a previous short circuit or heavy overload.

At the same time, determine the number of circuits in the panel; that is, how many single-pole (120-volt) circuit breakers or fuses are present? How many double-pole (240-volt) circuit breakers are present. A double-pole circuit breaker will look like two single-pole breakers with their handles joined with a metal or plaster connector. While looking inside the panelboard, determine if the panelboard cover is loose or missing. Also note any rust that you may find, along with any open knockouts. A knockout is a partially serrated round opening in the housing that may easily be knocked out with a screwdriver to accept cable connectors. If there are any openings in the panelboard housing, they should be closed with the appropriate fitting.

One of the most important safety factors of any electrical system is the "ground." You should be able to see a bare copper wire coming out of the panelboard and terminating at a ground clamp attached to a nearby water pipe. Sometimes there will be two ground wires; one terminating at a water pipe and the other terminating to a driven ground rod outside the building. If you can find no such ground wire, the system is dangerous and should be noted. Check to see if the ground clamps are tight and not severely corroded — both at the pipe and inside of the panelboard.

This is also a good place to determine the wiring method since this is where all the conductors in the home terminate. Note if the wiring is type NM cable, BX, etc., or a combination of these.

Branch Circuits And Feeders

Once the main service is out of the way, you will want to check the various outlets throughout the house. This includes all receptacles,

Electrical Systems

lighting outlets and their related wall switches. Note any outlet or junction boxes that do not have a cover or wall plate.

One sure way to tell if the rooms have sufficient outlets (if the home is occupied) is to note an excessive number of extension cords. If many exist, chances are there are not enough receptacles in the room. Also feel of several wall plates covering the receptacles. If they feel warm, or are discolored, they are either overloaded or have loose connections, or both.

Each light should be controlled by one or more wall switches; try each one to see that it functions properly. Also note flickering lights when a motor comes on; that is, a pump motor, washer, or garbage disposal. When any heating appliances are turned on, do the lights dim? This is an indication that the circuits are overloaded. If you see lights dim for no apparent reason, then the service neutral conductor is probably loose.

All duplex receptacles should be of the three-prong, grounding type. Use a receptacle testing to check a few receptacles to see if the grounding or bonding wire is in tact and doing the job for which it was intended.

Ground-Fault Circuit-Interrupters

The *NEC* states in Section 210-8 that certain residential outlets must be provided with ground-fault circuit-interrupters. This includes receptacle outlets installed in the following locations:

- Bathrooms, garages, outdoors, crawl spaces, unfinished basements and kitchen receptacles with 6 feet of the kitchen sink.

However, this code regulation has only been in effect for about the past decade or so. Therefore, you will find many homes, built prior to, say, 1960 that will have no ground-fault circuit-interrupters installed. They should still be recommended, but are not mandatory.

The checklist in Figure 10-17 should help you with a thorough inspection of the home's electrical system. Look this list over carefully.

INSPECTION CHECKLIST — ELECTRICAL					
Service Entrance:	120 V, two-wire		120/240 V three-wire		
Service Panel:	Fused panel		Circuit-breaker panel		
Service Size:	100 amp		150 amp	200 amp	
Number of Circuits:_____					
Wiring Method:	Type NM cable		Armored cable	Other	
				Yes	No
1.	Is the service panelboard in good condition?				
2.	Is the service panelboard readily accessible?				
3.	Is the service properly grounded?				
4.	Do living areas have sufficient receptacles?				
5.	Do all light switches operate?				
6.	Do all outlet boxes and junction boxes have covers?				
7.	Do all outside outlets have watertight covers?				
8.	Are ground-fault circuit-interrupters provided for the required outlets?				
9.	Are all duplex receptacles of the three-wire grounding type?				
10.	Do any of the outlets feel warm or are they discolored?				
11.	Does the wiring in general meet Code requirements?				
Additional features: _____					
Quality of installation: Good_____ Fair _____ Poor _____					
Repairs needed:		1.			
		2.			
		3.			
		4.			

Figure 10-17: Checklist for electrical systems.

Chapter 11
Heating, Ventilating and Air Conditioning

Few items in the home are more expensive, or are more important to the occupant's comfort, than the heating, ventilating, and air conditioning (HVAC) system. The condition and efficiency of an HVAC system also play an important role in the overall value of the house.

You don't have to be an engineer to know what to look for when inspecting a home's HVAC system. Just follow the tips in this chapter and you'll be well on your way to learning the accepted practices and the pitfalls that should be avoided.

Don't be fooled. HVAC systems have been vastly improved in the last 20 years, but too many systems have been installed according to dated standards. Tests made by a mobile field lab show that over half the homes in the United States have outmoded, inadequate or low-grade HVAC installations.

Don't assume that, because a house appears attractive and up-to-date, it has good heating or air conditioning. Don't assume that, because you see a name-brand piece of equipment, the house will be comfortable. The boiler and furnace are only one component of a complex, highly engineered system. That shiny name-brand furnace or condensing unit doesn't mean too much if it's undersized, or if it comes with incomplete controls, or with a sloppily designed and inadequate distribution system. Ductwork often costs more than the furnace. That's where a contractor can cut corners and it's here — not with the furnace — that most complaints about inadequate heating or air conditioning belong.

Don't be misled by signs or statements that the heating system is UL listed, or has some other testing agency's approval and conforms to local building codes. These standards are usually minimal. They don't mean the home has a good or economical system.

This chapter is designed to show you what to look for in a residential HVAC system. You will also learn how to test these systems to insure that they meet current standards set forth by the lending institutions hiring your services.

TYPES OF HEATING SYSTEMS

In general, heating systems for the home are divided into two types — central heating and individual space heating. These systems may be further subdivided into forced air, hot water, steam, convection, and solar heat. The fuels used for any of these systems include natural and LP gas, oil, electricity and such solid fuels as coal, coke, peat, and wood. The amount of heat is regulated by either central or zone controls.

Warm Air Systems

Warm air, at the present time, is the choice in three out of four new homes. Usually its cost is about 20% less than a hot water system of comparable quality.

Manufacturers of warm air heating equipment cite these other strong points:

- It delivers heat in a hurry and shuts off immediately when you don't want it. You don't don't have to sweat or shiver waiting for a boiler or a slab to adjust to the heat

Heating, Ventilating and Air Conditioning

requirement. This is especially important in houses with large glass areas.

- Forced warm air systems filter the air, eliminating much dust and pollen. If the homeowners are subject to airborne allergies or a sinus condition, this may be a deciding factor.

- Humidity is readily controlled. Lack of humidity may affect the respiratory system, rasp the throat, make people cough, and invite colds. It also dries out and damages pianos, furniture, and the house structure. With proper humidity you can feel comfortable at a lower temperature and thus save 3 to 4 percent on fuel bills.

- Air is kept moving, not stagnant. A slight but continual air circulation adds comfort. Without it a room becomes stuffy. A good system also brings in air from the outside, filters and mixes it with inside air so that room air remains fresh, never becomes stale. In summer, mere air movement even without cooling will make you more comfortable. With the burner shut down, the blower can be used to circulate filtered air.

- The addition or conversion to summer cooling is simpler and cheaper with a forced air system. The same ducts and blower may be used for both heating and cooling. The addition of cooling coils, refrigerant lines, and an outside compressor is all that is required to add on cooling to an existing forced air furnace. See one type of installation in Figure 11-1 on the next page.

Hot Water Heating

Hot water systems cost more, but they have long enjoyed a reputation for quality and have been found in the most expensive homes. Manufacturers cite these special advantages.

- It provides more even heat. Water retains its heat far longer than air, and sudden, wide fluctuations are seldom

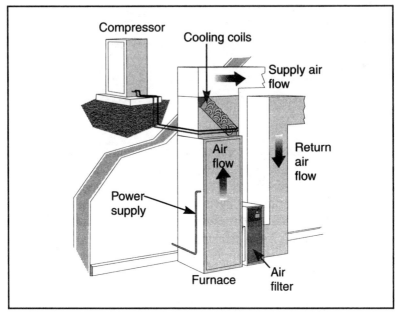

Figure 11-1: Forced-air furnaces allow easy cooling. Here cooling coils are located in the plenum and the condensing unit is located outside the building, with refrigerant lines connecting the two.

encountered. This heat-carrying power means a superior ability to reach far ends and corners of the house. Long and complicated ductwork is expensive, means heat loss and low efficiency.

- Boilers are more compact, can be placed in a minimum space. Boilers are made that are less than 36 inches high and 24 inches wide.

- Pipes are small and flexible. There are no bulky ducts to limit the occupants use of basement recreational areas, and with space costing dollars per square foot, this can mean a tremendous saving.

- The system is sealed. No smoke, dirt, odors, or possibly germs, can be conducted from room to room. Closing of doors doesn't affect heat distribution. Any air movement is gentle, not drafty. See Figure 11-2.

Heating, Ventilating and Air Conditioning

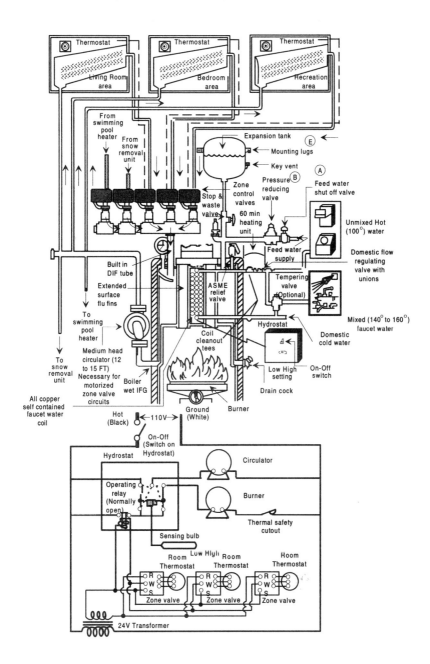

Figure 11-2: Modern zone-controlled hot-water heating system.

- The boiler may also be used to furnish domestic hot water, and save by not having to get a separate heater. The boiler can also be used to heat swimming pool water, or melt snow and ice on walks and driveway.

Electric Space Heating

The use of electric heating in homes has risen tremendously since 1960, and electric heating, at one time, showed all signs of becoming the principal type of heat for all types of structures. However, during the fuel shortage of the 1970s, electricity, along with other types of fuel, rose sharply in price, which made all of us take another look at more economical sources of heat. Still, electric heating has become very popular for use in residential occupancies, mainly because of the following advantages:

- Electric heat is noncombustible and therefore safer than combustible fuels.

- It requires no storage space, fuel tanks, or chimneys.

- It requires little maintenance.

- The initial installation cost is low.

- The amount of heat may be easily controlled since each room may be controlled separately with its own thermostat.

- It is predicted that electricity will be more plentiful than other fuels in the future.

The type of electric heating system used for a given application will usually depend on the structural conditions, the kind of area, and the purpose for which the area will be used. The owner's preference will also enter into the final decision.

Electric heating equipment is available in baseboard (Figure 11-3A), wall (Figure 11-3B), ceiling (Figure 11-3C), kick space (Figure 11-3D), and floor units; in resistance cable embedded in the ceiling or concrete floor; and in forced-air duct systems similar to conventional

Heating, Ventilating and Air Conditioning

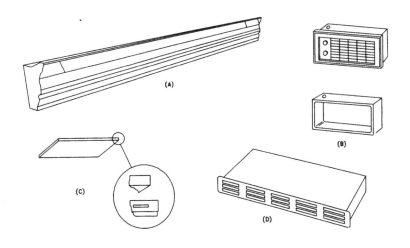

Figure 11-3: Various types of electric heating units: (A) baseboard unit, (B) wall unit, (C) ceiling unit, (D) kick-space unit.

oil- or gas-fired furnace hot-air systems except that the source of heat is electric heating elements.

In general, heating equipment should be located on the outside wall near the areas where the greatest heat loss will occur, such as under windows. The controls or wall-mounted thermostats should be located on an interior wall, about 50 inches above the floor, to sense the average temperature.

Electric boilers are also available for central hot-water heating, as well as electric furnaces and heat pumps for central air heat/cooling.

Miscellaneous Equipment

Combination Heating/Cooling Units: The advantages of through-wall heating and cooling units are similar to those of electric baseboard heaters except that they also have cooling capability.

Heat Pumps: A heat pump is a system in which refrigeration equipment takes heat from a source and transfers it to a conditioned space (when heating is desired), and removes heat from the space when cooling and dehumidification are desired.

Infrared Heaters: These are assemblies that make use of the heat output of infrared lamps or other sources. Such heaters provide fast response, high-temperature radiation and are particularly suited for use in locations in which it is difficult or impractical to maintain air temperatures at comfortable levels.

Fan-Driven Forced-Air Unit Heaters: The heat source may be provided by either electric resistance heating or piped steam or hot water. Cooling may also be used in combination with the heat when cold water from a chiller is circulated through the coils of the unit.

High-Velocity Heating and Cooling Systems: High-velocity systems have been used in commercial buildings for quite some time, but due to the noise once common to this type of system, very few were installed in residential buildings. Now, however, new designs in this type of system have reduced the noise level to the point where it is quite acceptable for residential applications.

The compactness of the equipment and the small ductwork and outlets make this system well suited for use in existing structures where the installation of conventional forced-air ducts would require too much cutting and patching. On the other hand, the small $3\frac{1}{2}$-inch-diameter ducts of a residential high-velocity system can be "fished" through wall partitions, corners of closets, and similar places. The system's small air outlets (only 2-inch openings) can be placed nearly anywhere and still provide good air distribution.

Solar Heat: A basic solar system consists of a solar (heat) collector that is usually arranged so that it faces south, a heat reflector mounted on the ground in front of the collector, a storage tank to hold the sun-heated water, a circulating pump, and piping. The operation of these components is simple, the storage tank is filled with water, which is pumped to the top of the heat collector. As the water flows over the collector, the sun heats it. At the bottom of the collector, the heated water is collected in a trough and then flows back to the storage tank.

The water may be pumped from the storage tank through a system of pipes to baseboard radiators in the living area; a heat exchanger in the water loop may transfer the heat to forced air for distribution via a duct system; or a heat pump may perform the exchange, using the water loop as both heat source and sink.

Radiant Heating Cable: Embedded heating cable is the most common large-area electric ceiling heating system in use today. The heating cable is laid out on a grid pattern and stapled to a layer of gypsum lath on the ceiling joists. Then the cable is covered with a layer of wet plaster, or another layer of gypsum board is placed over the cable. Heat cable is available on reels for easy payout as the grid is formed. The nonheating leads on the ends of the run of heating cable are brought up through the plates and fed down through the wall to the control thermostat or relay. Cables may be obtained in many lengths and spaced to satisfy room heating capacity requirements.

Miscellaneous Heaters: Since about 1973, much experimenting has been done with alternative heating sources and of these, the wood furnace seems to have gained the most popularity in certain areas. Modern wood furnaces are automatically controlled and most require filling only once each day. Their efficiency is high where a plentiful wood supply is available at reasonable cost — such as for rural residents with a woodlot — these systems are difficult to surpass when it comes to saving energy and heating expenses. Although usually more expensive to install than conventional models, wood furnaces will quickly pay for themselves in fuel savings if, as mentioned previously, a cheap source of wood is available.

BASIC AIR-CONDITIONING CONCEPTS

The fundamental concepts of air conditioning are not understood or even considered by the millions who enjoy the comfort heating and cooling produces. Nevertheless, it is readily accepted as part of the American scene. This is something of a phenomenon, so air conditioning requires definition:

> Air conditioning is defined as a process that heats, cools, cleans, and circulates air and controls its moisture content. Ideally, it performs all of these functions simultaneously and on a year-round basis.

Air conditioning makes it possible to change the condition of the air in an enclosed area. Because modern people spend most of their lives in enclosed areas, air conditioning is more important and can produce a greater sustained beneficial influence on humans than even outdoor weather. People work harder and more efficiently, play longer, and enjoy leisure more comfortably because of air conditioning. Scientific achievements and applications have been outstanding.

- Military centers, which track and intercept hostile missiles, are able to operate continuously only because air is maintained at suitable temperatures. Without air conditioning, the mechanical brains in these centers would cease to operate in a matter of minutes because of the intense self-generated heat.

- Atomic submarines can remain submerged almost indefinitely due, in part, to air conditioning.

- Modern medicines such as the Salk vaccine are prepared in scientifically controlled atmospheres.

- Human exploration of outer space has been greatly simplified by air conditioning.

Body Comfort

The normal temperature of the human body is 98.6 degrees F. This temperature is sometimes called subsurface or deep tissue temperature as opposed to skin or surface temperature. An understanding of the manner by which the body maintains this temperature will help in understanding the manner by which the air conditioning process helps to keep the body comfortable.

All food taken into the body contains heat in the form of calories. The "large" or "great" calorie, which is used to express the heat value of food, is the amount of heat required to raise one kilogram of water 1 degree C. As calories are taken into the body, they are converted into energy that is stored for future use. The conversion process generates heat. All body movements not only use up the stored energy, they also add to the heat generated by the conversion process.

For body comfort, all the heat produced must be given off by the body. Because the body consistently produces more heat than it requires, heat must constantly be given off or "removed." The constant removal of body heat takes place through three natural processes that usually occur simultaneously. These processes are *convection, radiation,* and *evaporation.*

Heating, Ventilating and Air Conditioning

Convection: The convection process of removing heat is based on two phenomena: (1) Heat flows from a hot to a cold surface. For example, heat flows from the body to surrounding air that is below body skin temperature. (2) Heat rises. This becomes evident when observing the smoke from a burning cigarette.

When these two phenomena are applied to the body process of removing heat, the following changes occur: (1) The body gives off heat to the cooler surrounding air. (2) The surrounding air becomes warm and moves upward. (3)As the warm air moves upward, more cool air takes its place, and the convection cycle is completed.

Radiation: Radiation is the process by which heat moves from a heat source (sun, fire, etc.) to an object by means of heat rays. This principle is based on the previously noted phenomenon that heat moves from a hot to a cold surface. Radiation takes place independent of convection, however, and does not require air movement to complete the heat transfer. It is not affected by air temperature either, although it is affected by the temperature of surrounding surfaces.

The body quickly experiences the effects of sun radiation when it moves from a shady to a sunny area. It again experiences radiation effects when the body surface closest to a fire becomes warm while more distant surfaces remain cool. Just as the heat from the sun and fire moves by radiation to a colder surface, the heat from the body moves to a colder surface.

Evaporation: Evaporation is the process by which moisture becomes vapor. As moisture evaporates from a warm surface, it removes heat and thus cools the surface. This process takes place constantly on the body's surface. Moisture is given off through the pores of the skin ad, as the moisture evaporates, it removes heat from the body.

Perspiration, which appears as drops of moisture on the body, indicates that the body is producing more heat than can be removed by convection, radiation, and normal evaporation.

Conditions That Affect Body Heat

- Temperature.

- Cool air increases the rate of convection; warm air slows it down.

- Cool air lowers the temperature of surrounding surfaces and, therefore, increases the rate of radiation; warm air raises the surrounding surface temperature and, therefore, decreases the radiation rate.

Humidity

Moisture in the air is measured in terms of humidity. For example, 50 percent relative humidity means that the air contains one-half the amount of moisture that it is capable of holding. To simplify the measurement of humidity, a unit called a *grain of water vapor* is used. A grain is a small amount; in fact, there are approximately 2,800 grains in one cup of water and 7,000 grains in one pound of water.

The following puts this information into practical use: Assume that a room has a temperature of 70 degrees F and four grains of water vapor for each cubic foot of space. If the room temperature remains at 70 degrees F and water vapor is added, the air in the room eventually reaches the point at which it cannot absorb more water. At this point, the air is saturated, and one cubic foot of room space now holds eight grains of water vapor. At 70 degrees F, 8 grains per cubic foot represents 100 percent relative humidity. The original room condition of 4 grains at 70 degrees F represents 50 percent relative humidity: 4 grains/8 grains = 0.50, or 50 percent.

Relative humidity, then, is obtained by dividing the actual number of grains of moisture present in a cubic foot of room air at a given temperature by the maximum number of grains that the cubic foot of air can hold when it is saturated.

The relative humidity changes when the temperature changes. For example, at 80 degrees F, the relative humidity is 4/11 = 0.37, or 37 percent.

If, instead of increasing the temperature to 80 degrees F, the actual moisture content of the air is decreased from 4 grains to 3 grains per cubic foot at 70 degrees F, the relative humidity is 3/8, or 37 percent again.

From the preceding examples, the means of changing relative humidity should become evident.

- To increase relative humidity, increase the actual moisture content of the air or decrease the air temperature.

- To decrease relative humidity, decrease the actual moisture content of the air or increase the air temperature.

A low relative humidity permits heat to be given off from the body by evaporation. This occurs because the air at low humidity is relatively "dry" and thus can readily absorb moisture. A high relative humidity has the opposite effect: It slows down the evaporation process and thus decreases the speed at which heat can be removed by evaporation. An acceptable comfort range for the human body is 72 degrees F to 80 degrees F at 45 percent to 50 percent relative humidity.

Air Movement: Another factor that affects the ability of the body to give off heat is the movement of air around the body. As air movement increases the following changes occur:

- The evaporation process of removing body heat speeds up because moisture in the air near the body is carried away at a faster rate.

- The convection process increases because the layer of warm air surrounding the body is carried away more rapidly.

- The radiation process tends to accelerate because the heat on the surrounding surfaces is removed at a faster rate, causing heat to radiate from the body at a faster rate.

As air movement decreases, the evaporation, convection, and radiation processes decrease.

TYPICAL AIR CYCLE

Indoor air can be too cold, too hot, too wet, too dry, too drafty, and too still. These conditions are changed by "treating" the air. Cold air is heated, hot air is cooled, moisture is added to dry air . . . removed from damp air, and fans are used to create adequate air movement. Each of these "treatments" are provided in the air-conditioning air cycle. See Figure 11-4.

The description begins with the fan because it is the one piece of equipment that starts the air through the cycle. A fan forces air into ductwork connected to openings in the room. These openings are commonly called outlets or terminals. The ductwork directs the air to the room through the outlets. The air enters the room and either heats or cools as required. Dust particles from the room enter the air stream and are carried along with it.

Air then flows from the room through a second outlet (sometimes called the return outlet) and enters the return ductwork, where dust particles are removed by a filter. After the air is cleaned, it is either heated or cooled depending on the condition in the room. If cool air is required, the air is passed over the surface of a cooling coil; if warm air is required, the air is passed through a combustion chamber or over the surface of a heating coil. Finally, the treated air flows back to the fan, and the cycle is completed.

Thus, the major parts of equipment in the air-conditioning cycle are the fan, supply ducts, supply outlets, space to be conditioned, return outlets, return ducts, filter, heating chamber, and cooling coil.

Fan: The principal job of the fan is to move air to and from a room. In an air-conditioning system, the air that the fan moves is made up of:

- All outdoor air

- All indoor or room air (this is also called recirculated air)

- A combination of outdoor and indoor air

The fan can "pull" air exclusively from outdoors or from the room, but in most systems, it pulls air from both sources at the same time.

Because drafts in the room cause discomfort, and poor air movement slows the body heat rejection process, the amount of air supplied by the fan must be regulated. This regulation is done by choosing a fan that can deliver the correct amount of air and by controlling the speed of the fan so that the air stream in the room provides good circulation without causing drafts. Of course, the fan is only one of the pieces of equipment that contributes to body comfort; others, such as supply and

Heating, Ventilating and Air Conditioning

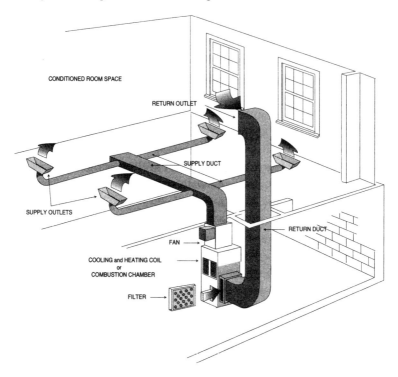

Figure 11-4: Major parts of an air-conditioning system.

return room outlets and cooling and heating equipment, are described in subsequent paragraphs.

Supply Duct: The supply duct directs the air from the fan to the room. It should be as short as possible and have a minimum number of turns so the air can flow freely.

Supply Outlets: Supply outlets help to distribute the air evenly in a room. Some outlets "fan" the air, others direct it in a jet stream, still others can do a combination of both. Because supply outlets can either fan or jet the air stream, they are able to exert some control on the direction of the air delivered by the fan. This directional control combined with the location and the number of outlets in the room contributes a great deal to the comfort or discomfort effect of the air pattern.

Room Space: The room is one of the most important parts of the air cycle description. The dictionary states that a room is an "enclosed space" set apart by partitions. If this enclosed space did not exist, it would be impossible to complete the air cycle because conditioned air from the supply outlets would flow into the atmosphere. The enclosed space, therefore, is all important. In fact, the material and the quality of workmanship used to enclose the space are also important because they help to control the loss of heat or cold that is confined in it.

Return Outlets: Return outlets are openings in the room surface that are used to allow room air to enter the return duct. They are usually located at the opposite extreme of a wall or room from the supply outlet. For example, if the supply duct is on the ceiling or on the wall near the ceiling, the return duct may be located on the floor or on the wall near the floor. This is not true in all cases, however, because some systems have both supply and return outlets near the floor or near the ceiling.

Keep in mind that the main function of the return outlet is to allow air to pass from the room.

Filters: Filters are usually located at some point in the return air duct. They are made of many materials, from spun glass to composition plastic. Other types operate on the electrostatic principle and actually attract and capture dust and dirt particles through the use of electricity.

The end purpose of all filters is to clean the air by removing dust and dirt particles.

Cooling Coil and Heating Coil or Combustion Chamber: The cooling coil and the heating coil or combustion chamber can be located either ahead of or after the fan but should always be located *after* the filter. A filter ahead of the coil is necessary to prevent excessive dirt, dust, and dirt particles from covering the coil's surface. A separate condensing unit (not shown), located on the outside of the building is also required for most types of air-conditioning systems.

All of the material described in this chapter, however, cannot be accomplished without some means of control, if only to stop and start a system. For example, without a controlling mechanism, an air-conditioning system would be turned on and run year-round. Even when the space became cool or hot enough, the system would continue to run and make the area more uncomfortable than if it had not been installed in the first place. Therefore, control is an important area of comfort conditioning for buildings.

TYPES OF CONTROLS

All heating, ventilating, and air conditioning (HVAC) systems, regardless of their size or quality, must be controlled, if only to turn the system off and on. The "control" may range from a simple toggle switch built into an electric heater to highly sophisticated computerized controls for an entire building or plant. In all cases, however, HVAC controls will be either mechanical, electrical, electronic, or pneumatic.

In general, a fully automatic HVAC control system will perform the following functions:

- The sensing element measures changes in temperature, pressure, and humidity.

- The control mechanism translates the changes into energy that can be used by motors, valves, relays, and the like.

- The connecting electric wiring, pneumatic piping, and mechanical linkages transmit the energy to the motor, valve, or other device.

- The motors or valves use the energy to operate toward some corrective action. Motors operate compressors, fans, dampers, and similar devices. Valves control the flow of gas to burners or to cooling coils and permit the flow of air in pneumatic systems. Valves also control the flow of liquids, such as water in a water-cooled condenser.

- The sensing elements in the control detect the change in conditions and signal the control mechanism or connecting means.

- The control stops the motor or closes the valve, terminating the call for corrective action. This action by the control prevents overcorrecting.

Control Action

Several types of actions are involved in the control of various HVAC systems. Most, however, are "two-position" controls—either simple or timed.

Two-position control permits the final control element, such as a relay, motor, valve, or similar device, to occupy one or the other of two positions. Examples would be to start or stop (two functions) motors for fans, compressors, pumps, etc.; or to open and close a valve (two function) in an air-conditioning system or a gas-fired furnace.

In simple two-position control, the control never catches up to the controlled condition. Rather, it corrects a condition that has just passed, not the condition that is taking place or about to take place.

Timed Two-Position Control: The ideal method of heating or cooling a space is by replacing heat lost or removing heat gained in exactly the required amounts. To do this, the control must respond to a gradual or average change in the space air temperature (controlled variable). An average change is produced by adding a tiny heater (timing device) near the thermostat temperature-sensing element.

As long as the air temperature at the thermostat is maintained within certain limits, the thermal heat cools to its "on" point and energizes the heating element; it then heats to its "off" point and de-energizes the heating element. It again cools to the "on" point and repeats the cycle.

The timer sequence for cooling the space is the opposite of the sequence for heating. The heater element turns off when the air at the thermostat is cool and turns on when the air is warm. When the element turns on, it adds heat at the thermostat element and causes the thermostat to call for cooling sooner than it would if the heater were not used. This causes cooling to be added to the space immediately and prevents the air temperature from rising as far as it would without the timer.

The timer or thermal heater principle reduces the "swing" or differential in air temperature in a room to the point at which the temperature remains almost constant. The differential is much greater when the timer is not used.

Summary of HVAC Controls

- HVAC controls can be electric, pneumatic, or electronic.

- Set point is the temperature, pressure, or humidity at which the control indicator is set.

- Control point is the temperature, pressure, or humidity recorded by the control.

- Deviation is the momentary difference between control point and set point.

- Corrective action is action taken to maintain the control point in reasonable agreement with the set point.

- Differential gap is the range through which a controller travels from the point at which its contacts open to the point at which they close.

- Offset is the difference between set point and control point.

- Primary element is that part of a control that is first to use the energy resulting from a change in the controlled medium.

- Primary elements can be bimetal, bellow, resistance wire, hair, wood, leather, gold strips, and the like.

- A fully automatic control system is one that senses a change in conditions and takes action to correct the change.

- Control action is "simple" and "timed" two-position.

- Electric controls use electric energy to transmit signals.
  Low voltage is 32 V or less.
  Line voltage is 120 V to 240 V.

- Control circuits can be three-wire, low-voltage or three-wire line-voltage; or two-wire, low-voltage or two-wire, line-voltage.

- Electric control circuits can be used on all types of heating or cooling equipment.

INSPECTING THE HVAC SYSTEM

The home inspector will probably encounter more forced-air heating systems that any other type. The heating chamber for a forced-air system is called a furnace, and can be "fired" by either oil, gas, or electricity. The heated air is distributed throughout the home by means of a fan that pushed the heated air through ductwork, called *supply duct*. This air enters each room at outlets which are referred to as supply *grilles, diffusers,* or *registers*. The "used" heated air is then returned to the furnace for reheating through return-air ducts — this time "pulled" by the same fan. The entire system is controlled by a thermostat to enabled the system to be energized when heat is needed.

Forced-air furnaces are manufactured in various grades from those guaranteed for 10 years or more, down to those guaranteed for only a year. The better grade furnaces have a thick, well-made heating chamber, and also a belt-driven fan with a pulley, similar to an automobile's fan belt. The cheaper units utilize direct-drive blowers; that is, a fan is installed directly on the motor shaft. The furnace fan, also called fan-coil unit, may easily be inspected at the furnace by removing the fan-coil unit cover. Most of the time this cover is hinged with a latch; others may require the removal of two or more screws. Once open, you should be able to readily tell if the fan is belt-driven or of the direct-drive type. The 10-year guarantee and belt-driven fan-coil unit are the major distinguishing features of a high-grade furnace.

The air filter should also be located in the fan-coil unit. While the fan-coil unit cover is off, remove the air filter and check it for cleanliness. A clean air filter can tell you if the existing heating system was well maintained. If extremely dirty, you can expect that the system has had little maintenance and would be reason to inspect further.

The design and installation of the air-duct system are essential to good heating. It must be designed so that the sufficient heated air is moved at the proper velocity to the spaces to be heated, yet not so fast as to be noisy. The most efficient duct design is called *perimeter duct*

Heating, Ventilating and Air Conditioning

distribution as shown in the floor plan in Figure 11-5 on the next page. Note that the supply outlets are located around the exterior walls (the house perimeter), while the return-air grilles are located on the interior walls of the house. This way, warm air is supplied in each room at the source of the greatest cold, and then moves across the room to the return-air grille — always providing even circulation of air in the area, which provides the greatest comfort.

When inspecting a forced-air heating system, there should be at least one warm-air outlet in each room for each exposed wall. Many systems, however, will use only one because the installation is cheaper. When you do find a house with a floor or ceiling diffuser on each exposed wall, you can almost rest assured that the heating system was installed properly.

Many furnaces in older homes that originally burned coal has been converted to oil or gas. Look on the furnace nameplate to see if the fire chamber is made of cast iron or steel. A cast iron one will usually last longer. Look inside the fire chamber for signs of cracking and around the exterior base for rust and deterioration.

If the home is less than 20 years old, it probably has the original heating equipment, installed at the time the house was constructed. If the house is more than 20 years old, the house may have had the old system upgraded, or else an entire heating plant has been installed. If not, chances are a new system will be required in the near future.

All heating systems need periodic inspection, cleaning and adjustments. As a general rule, gas and oil systems should be tuned-up every year before the start of the heating season. Doing so will cut fuel costs and also prolong the life of the equipment. Electric furnaces do not require a "tune-up," but they should be cleaned at least once a year. Some of the moving components may also need to be lubricated according to the manufacturer's instructions. If a professional does this, a service record will normally be attached some place on the furnace, much the same as a garage attaches a maintenance ticket to the inside of car doors when an oil change is made. Look for these service records when inspecting the furnace; much can be learned about the heating system by reviewing these. For example, if the heating system was serviced at least once each year, the system was maintained as it should be. However, if the service record indicates infrequent service calls, and then only when the system didn't work properly, the system is a good candidate for future problems.

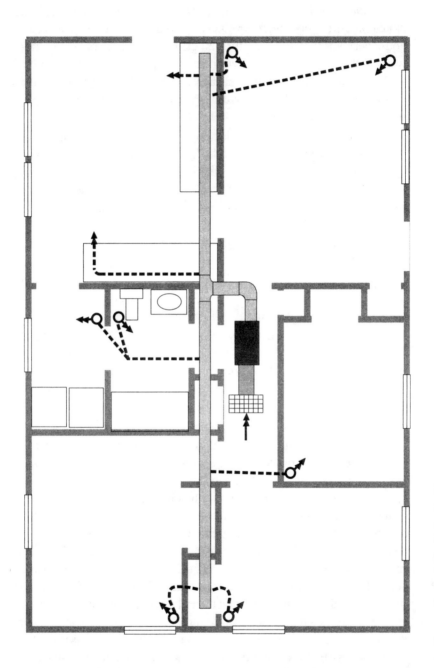

Figure 11-5: Floor plan of a home, showing the ductwork.

Heating, Ventilating and Air Conditioning

Sizing the heating system is also very important. One that is too small will run almost constantly on a cold day, trying to bring the system up to the desired temperature, yet will never quite make it. One that is too large will shut off and on frequently, wasting fuel.

Heating units are normally rated in Btu's (British thermal units). Sizing the furnace or other heating system requires careful calculations and the consideration of several factors, but you can make an approximate calculation once you have determined the area of the home. In areas where the outside design temperature will be as low as 0° F, while the inside temperature is to be maintained at 70° F, you can multiply the area (in feet) by 45 (Btu's) and come up with an approximate size of the heating unit for the average home.

For example, if the total area of the house is, say, 1650 sq. ft., you would multiply this figure by 45 for a total heat loss of:

$$1650 \times 45 = 74{,}350 \text{ Btu's}$$

The size of the heating plant can usually be found on the nameplate of the heating unit. Once you have made your calculation, see how it compares to the nameplate rating of the heating unit. If it is considerably higher or lower than your calculation, further investigation may be necessary. However, be aware that this is a very rough means to check the correct size of the heat loss. It should not be used to size an actual installation. There are also many factors that can change this figure significantly: larger glass areas than usual; inadequate insulation; higher-than-normal ceilings, and similar items. So don't take this equation as gospel.

Combination Units

The better homes built within the past couple of decades will have both heating and cooling in the same central unit. A good air-conditioning system must do the following:

- Heat the air and add humidity in winter.

- Cool the air and remove humidity in summer.

- Circulate and filter the air.

An efficient heating\cooling system takes good planning, and good workmanship, and good equipment to meet all of the above requirements.

The ideal system maintains constant daily temperature in the house. It should circulate air without drafts. Temperatures during the day should not vary more than 2 or 3 degrees F in winter, and temperatures between rooms, unless a zoned system, should not vary more than 4 to 5 degrees F. When outdoor temperatures are below freezing, there should not be more than a 5 degree difference between floor and ceiling air temperatures.

Even if the house you are inspecting is vacant, the heating system should already be turned on to maintain a temperature of about 40° F to keep water pipes from freezing and also to keep building materials from cracking. Consequently, if you are inspecting such a home on a cold day, turn up the thermostat to 75° F and listen for the heating plant to come on. If the house has a forced-air system, the temperature in the house should rise 10 to 15 degrees in 15 minutes. If it's a hot-water system, it may take a little longer. If the temperature falls short of this, something is amiss.

A similar procedure may be used with a cooling system. A vacant house on a hot day may have temperatures exceeding 90° F. If inspecting such a home, first open the doors and a few windows to let out some of the hot air in the house. Then activate the air-conditioning system. The house should cool as much as 15° F within 30 minutes.

Electric Heaters

Individual electric baseboard heaters may be checked any time of the year. Merely go through the house and turn the thermostat up as high as it will go in each room. By the time you make the "circle," the first heater that you turned on should be warm. If so, turn the thermostat off (or to where it was originally if inspection is made during cold weather), then go to the next room and do the same thing, and so on until all heaters have been checked.

If a forced-air electric furnace is used, turn the thermostat up all the way and listen for the blower to come on in 5 to 10 minutes. When it comes on, place your hand near a supply outlet and see if you feel warm air. After the check, adjust the thermostat to its original position.

Heating, Ventilating and Air Conditioning

A checklist in Figure 11-6 on the next page will help you with the inspection procedures for the home's HVAC system. You will also want to modify this as experience dictates.

INSPECTION CHECKLIST — HVAC SYSTEMS					
Type of heating systems:	Forced-air	Hot water	Electric baseboard	Other	
Type of fuel:	Oil	Gas	Electric	Other	
Type of cooling system	Forced-air	Chilled water	None		

		Yes	No
1.	Do all rooms have air-supply outlets or heating units?	___	___
2.	Is all heating/cooling equipment in good condition?	___	___
3.	Are air-supply outlets (openings) adjustable?	___	___
4.	Does the duct system have an air filter?	___	___
5.	Is the heating/cooling unit sized correctly?	___	___
6.	Is the heating/cooling unit of high quality?	___	___
7.	Are there any cracks in the furnace fire chamber?	___	___
8.	Do all air ducts have dampers?	___	___
9.	Do all hot-water radiators have shutoff valves?	___	___
10.	Are all ductwork joints properly sealed?	___	___
11.	Has the HVAC system had a recent checkup or maintenance service?	___	___
12.	Is the thermostat located properly and in good condition?	___	___
13.	Have the cooling coils been charged with refrigerant recently?	___	___
14.	Is all ductwork insulated?	___	___

Additional features:

Repairs needed:

Quality of installation: Good____ Fair____ Poor____

Figure 11-6: Inspection checklist for HVAC systems.

Chapter 12
Plumbing Systems

The home plumbing system has two main purposes: to bring in a water supply, and then dispose of it, along with any other domestic waste. Both the water supply and disposal systems usually are made up of several hundred feet of pipe and related fittings to join the pipes and to make turns or bends in the pipe runs.

The main supply of water coming into the house and the distribution pipes that carry this water is under pressure — between 30 and 60 pounds per square inch (psi). Consequently, both the cold- and hot-water supply pipes can be relatively small in diameter, yet supply sufficient water to the various outlets or faucets.

Water and other wastes in the disposal part of the home's plumbing system always flows by gravity. For this reason its pipes and fittings must be larger in diameter to carry the required flow without clogging or backing up. Both systems — the supply and disposal — when installed properly, are designed to operate safely and quietly.

The home inspector is required to check the various parts of the home's plumbing system to ensure that it operates properly, and more important, that the system is safe. If contaminated water should get into the drinking water, it can sicken or kill. By the same token, an improperly installed drain trap can allow poisonous sewer gas back into the house; this too, can sicken or kill.

This chapter is designed to provide the home inspector with a good knowledge of the various parts of the typical residential plumbing system, how each functions, and how to inspect and report on these systems.

THE BASIC SYSTEM

The diagram in Figure 12-1 shows a pictorial diagram of a typical residential plumbing system. Note that branches are taken off the main supply at convenient places and led to the plumbing fixtures through water pipes. The size of pipe, grade or slope of the feed lines, alignment, and tight joints and connections are essential considerations in a good water distribution system. The distance that water has to travel to the outlet is very important in determining the size of pipe and the grade. If the pipe is sized too small, the water will travel at a higher velocity and become noisy and aggravating to those living in the house. Correct pipe sizing also keeps the water flow uniform when water to another plumbing fixture is turned on or off. This way, no one fixture can rob the water intended for another.

The inside walls of all pipe produce a surprising amount of frictional drag and for this reason, the pipe should be as short as possible and have as few connections and bends as possible. All distribution pipes should be securely fastened to the joists and wall supports to prevent vibration of the service lines.

The supply mains should be graded to one low point in the basement so that a drain cock will allow complete drainage of the entire system. Any part of the piping that cannot be so drained must be equipped with a separate drain cock or opening. In general, a pitch of $\frac{1}{4}$ inch to the foot of pipe is sufficient to allow complete drainage of fresh water. All the fixture traps can be drained individually.

Shut-off valves are installed at logical and accessible places for stopping the flow of water to any fixture or part of the distribution system that is giving trouble. In this way, repairs can be made to one fixture or faucet without inconvenience to the rest of the house. Furthermore, trouble can be quickly stopped by simply closing the appropriate shut-off valve as required.

The main cold water valve is located in the basement or near the water main entrance to the house and on the house side of the meter.

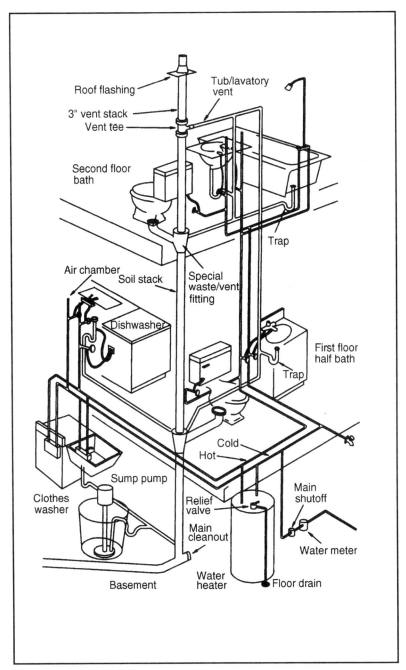

Figure 12-1: Typical home plumbing system showing both the supply and disposal systems.

This valve controls the flow of water. The hot water tank should have its own shut-off valve so that the hot water system can be closed and controlled without affecting the rest of the system.

Drip cocks should be installed on the riser side of each valve in the system so that the closed sections can be drained of standing water. If possible, the drip cocks should be placed at the base of every riser (a pipe that runs in a vertical direction).

Insulation of the cold water pipes within the house is very often neglected by the plumber or omitted by the builder in the interest of economy. This is a poor economy, however, for the home owner because such insulation will prevent ceiling and wall damage from wet pipes. During warm and humid weather the cold water pipes will collect moisture on the outside of the pipe because the water within the pipe is cooler than the outside air.

In locating the cold water distribution pipes to the fixture risers, it must be remembered that all cold water outlets on standard fixtures are located on the left side as the user stands facing the fixture.

Air chambers consist of pipes a foot or so long and capped on the upper ends. Under pressure, water rises in the air chambers so formed, compressing the trapped air at its top. The purpose of air chambers is to cushion the shock of fast running water through the water supply system, as it is stopped abruptly when the faucet is turned off.

Quick-acting solenoid water valves on washing machines and dishwashers turn off quite abruptly. Then, internal water supply system pressures can reach 500 psi and more, kicking out in all directions on the piping. Overpressures like these create a sound called water hammer, which is familiar to most owners of old houses. They also endanger the soundness of the water supply system. Air chambers do away with water hammer and the overpressures caused by it.

No faucet or water-using appliance should be without an air chamber on both the hot and cold supply.

Drain-Waste-Vent System

The drain-waste-vent, or simply DWV, system begins at the fixtures and appliances where the water supply system stops. Since drained water is not under pressure, the DWV piping must slope slightly toward where the waste water is supposed to go by gravity flow.

Plumbing Systems

DWV fittings are designed with slightly-less-than-square (90°) turns to help the plumber maintain slope. Water flows from each fixture and appliance to the city sewer or to a private sewage disposal system in the yard. The DWV piping becomes even larger as the collected wastes from more and more fixtures and appliances flow into it.

The first DWV pipe that leads waste away from a fixture is called a *waste* pipe. Toilets require large outflow pipes because of the great flow of water coming from them and the solids that must be carried along without clogging. Because of this, toilet drainage pipes are not called waste pipes, but rather are called *soil pipes.*

Stacks: Fixture waste pipes and toilets usually empty into what is called a stack, or soil stack. A stack is a vertical pipe that is open above the roof (vent through roof). At its lower end, a stack leads into the building drain. Waste water flows to the bottom of the stack and into the building drain. A typical drainage system is shown in Figure 12-2.

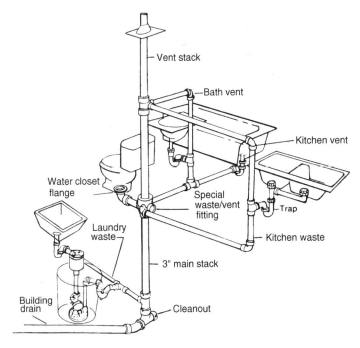

Figure 12-2: Drain waste and vent system.

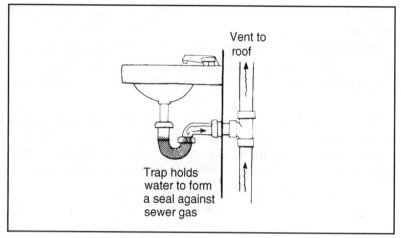

Figure 12-3: Operating principles of a plumbing trap.

Traps: A trap is a U-shaped pipe that will allow water and waste to pass through, but prevents gases and vermin from entering the DWV system. Every appliance and every fixture must have a trap. See Figure 12-3.

Water closets contain built-in traps because their intricate bowl passages are trap-shaped. The water you see in the bowl is part of a toilet's trap seal. In fact, when a toilet finishes flushing, the trap's siphon action often depletes most of its trap-sealing water. For this reason, a toilet tank's fill mechanism is designed to replenish the lost trap-sealing water as the tank refills. See Figure 12-4.

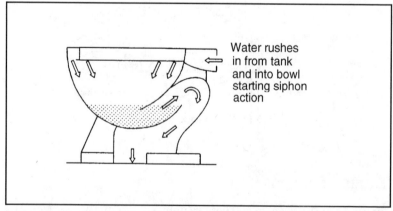

Figure 12-4: Operating principles of a water closet.

Plumbing Systems

Other fixtures and appliances use separate traps. Traps for sinks and lavatories are either P-shaped or S-shaped, but both contain a U-shaped trap section. These traps hide underneath fixtures. One end of the trap connects to a tailpiece coming out of the fixture drain. The tap must be at least as large as the tailpiece. The other end slips into the DWV system's waste pipe. Slip-nut connections with soft ring-type gaskets make slip joints both water- and gas-tight.

Tub and shower traps may be drum-type (more costly) or else large-sized P-traps. They are located in or beneath the floor under the fixture.

Wherever a trap is located, access must be provided for cleaning should it become clogged. Toilet and sink-lavatory traps can be cleaned from above, through the fixture drain. Some bathtub and shower traps are accessible for cleaning through the drain; some aren't. These are designed for cleaning from above through a hole in the floor or from below in the basement or crawlspace.

The best P- and S-traps are built with a cleanout opening at the lowest point of the dip. Removal of the cleanout plug allows draining and direct access for cleaning.

Vents: The necessity for traps brings on still another plumbing necessity — venting. Water rushing along by gravity through a pipe creates a suction or vacuum at the high end of the pipe above it. This is called siphon action. Siphon action is powerful enough to suck all the water out of a trap and leave it nearly dry, as happens in a toilet bowl after a flush. But, since other fixtures and appliances are not designed to replace siphoned-off trap water as a toilet is, some means of preventing trap siphoning must be built into the DWV system. This is accomplished by venting the system to outside air. See Figure 12-5 and 12-6 on the next page. Venting inside the house would work too, but then sewer gasses would escape into the house. Thus, venting is done outside above the roof.

Venting also prevents any pressurization in the DWV system or sewer system from building to the point where it could force past a trap's water seal. Every trap in a plumbing system should be vented.

Cleanouts: The purpose of a cleanout is to permit the removal of stoppages in the system caused by grease buildups or other problems associated with draining waste by gravity flow. Every horizontal drainage run, but not every vent run, must have cleaning access. This access is often through a cleanout opening at the higher end of the run.

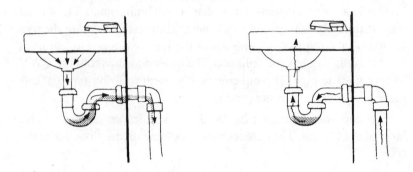

Figure 12-5: With no vent, trap water siphons off, leaving too little in trap to stop sewer gases.

The sewer line leading away from the house also needs an accessible cleanout opening for rooting it out.

The sewer line leading away from the house also needs an accessible cleanout opening for rooting it out, should it become blocked.

DWV Fittings

Drainage fittings are designed with gentle bends and curving inner surfaces that reduce resistance to flow. This also helps solids to pass through the fittings more easily.

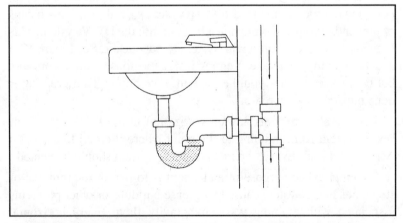

Figure 12-6: With vent, air rushes in to prevent siphoning of trap, and gas seal remains intact.

The common DWV-type fittings are elbows, tees, wyes and couplings. Elbows are manufactured as 22½-degree, 45-degree, 60-degree and 90-degree. Tees make a 90-degree angle; wyes make a 45-degree angle. DWV fittings, too, offer reducers, adapters and other special fittings for an easier, better job. One of these is the closet (toilet) flange, that couples the bottom of the toilet effectively to the DWV system.

Sewer-Septic System

In most towns and cities throughout the United States, public sewer systems are provided to carry away waste from buildings. This waste flows (by both gravity and pumps) to a sewage treatment plant where it is treated and then disposed of — usually in a nearby river or ocean.

Houses that you inspect outside of a town or city, however, will probably have a private sewage disposal facility called a septic system. A system consists of a sewer line running from the house, a septic tank and a drain field. The septic tank is where solid and liquid household wastes decompose by bacterial action. Solids settle out or flat while the resultant liquid, called effluent, flows out into the drain field. There it is absorbed into the ground.

The entire septic system is installed below the ground. All piping slopes from the house toward the far end of the seepage field for low gravity flow. Solid pipes are used for the house sewer, and from the septic tank to the distribution box; perforated pipe is used for the drain fields. See Figure 12-7.

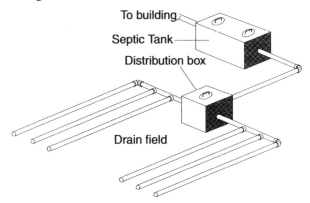

Figure 12-7: Typical septic system, used for rural waste disposal.

HOT WATER SYSTEMS

The typical electric water heater consists of a tank, electrical heating elements, thermostats to control the heating elements, and a safety valve. In operation, cold water enters the tank from the domestic cold-water line and is then heated by one or two heating elements. During this process, the heated water rises to the top of the tank where it is discharged through the hot-water outlet pipe. Thermostats are utilized to sense the temperature of the water and to control the heating elements to maintain the temperature of the water leaving the tank for use in the home.

There are several types of electric water heaters. The most conventional is the tall, round style, such as the one in Figure 12-8. Another is the cabinet type, which fits in with modern kitchen cabinets, offering a pleasing appearance while providing additional kitchen work surface.

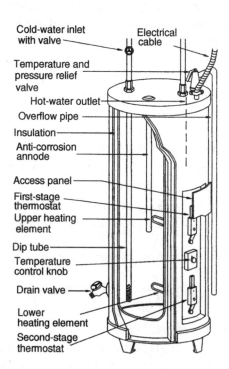

Figure 12-8: Components of a typical electric water heater.

Water heaters are normally classified according to the capacity of the water tank. Electric water heaters, for example, range in size from small units of 10- to 15-gallon capacity up to large units of 150 or more gallons, but the average size will run between 30 and 82 gallons. Some of the smaller tanks may be used on 120-volt circuits, but the majority of electric water heater circuits are designed for use on single-phase, 240-volt circuits.

Two types of water-heating elements are in general use: the immersion heater and the external element. Of these two types, the immersion type is by far the most common. With this type of heater, the heating element is installed through an opening in the tank wall so that it is in direct contact with the water in the tank. The external element straps around the tank like a belt and heats the water through the wall of the tank. With this latter method, the tank itself must first be heated and then the heat from the metal tank is transferred to the water. The only practical use for external elements is when a gas, wood, coal, or oil-fired water heater is being converted to an electric type. The external elements may be strapped to the outside of the tank, wired using external thermostats, and the heater is ready for use. The old means of heat (coal, oil, gas, etc.) may then be disregarded. However, such a conversion must be done by a qualified expert to ensure that the installation is done in accordance with national and local codes.

Two general types of thermostats are also in use. The immersion type extends inside the heater tank, similar to immersion-type heating elements. The external type is mounted flush against the outer surface of the tank. Of these two types, the external type is the most commonly used.

Two factors contribute to the amount of hot water that a given water heating system can produce in a given period of time:

- The wattage rating of the heating element.

- The capacity of the tank.

If the consumer's need for hot water is small, such as hot water for only a half-bath or an office rest room, usually a small tank with a low-wattage heating element is all that is needed. If the need for hot water is large, then a larger storage tank with high-wattage heating elements will be needed.

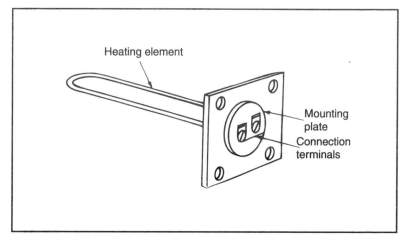

Figure 12-9: Immersion-type heating element for residential water heater.

Electric Water-Heating Elements

A typical immersion heating element is shown in Figure 12-9. It is very similar to the tube-type elements used on some types of electric hot plates and electric ranges. In general, the units consist of a nichrome element, which is first imbedded in a refractory material and then encased in a metal tube. The refractory material is a good electrical insulator and conductor of heat. This material then protects the element from a ground fault or from shorting out, yet readily conducts the heat generated to the outer metal tube so that it is able to heat the surrounding water.

The metal tube is watertight to protect the heating element from direct contact with the water, which in turn helps prevent a short circuit between the element and the grounded metal water tank. This metal sheath is usually made of copper tubing, which is brazed to a mounting flange. This flange is sometimes threaded and screws directly into threaded openings in the tank; in other cases, the flange is provided with bolt holes for direct bolting to the tank wall. In both cases, a watertight fit is provided. Since the immersion-type element operates in direct contact with the water, it is generally more efficient than the external types.

The external elements consist of flexible belt-shaped units that are strapped tightly around the outside of the tank wall. The heating element is a continuous coil of resistance wire that is sealed in an insulating material such as asbestos to prevent it from causing a ground fault between it and the metal tank walls.

In operation, electric current flows through the resistance wire, causing it to heat up and transfer its heat to the metal wall of the water tank. The heat from the wall, in turn, is transferred to the water in the tank, causing it to heat to the required temperature. It should be evident that external elements are not as efficient as the internal type; since the external type is located outside the tank, some of the heat never reaches the water inside the tank.

Thermostats

Thermostats are used in conjunction with the heating elements to control the current to the heating elements, which in turn regulates the water temperature. Most are adjustable for the desired water temperature from 100 to 160 degrees F, with 140 degrees F being the most common setting for residential use. However, when automatic dishwashers are used, a setting of 160 to 170 degrees is recommended. On most water heaters, two thermostats will be used, one for the top heating element and one for the bottom.

Water-Heating Systems

To obtain the best results and highest efficiency from water-heater installations, certain piping recommendations should be followed. In general, the water heater should be located and the piping installed so that the least amount of heat loss occurs. Factors governing the amount of heat loss are:

- The length of the hot-water lines.

- The size, weight, and type of pipe.

- Amount of pipe insulation.

Hot-water lines should be kept as short as possible as the water standing in the pipe between uses begins cooling to the ambient or surrounding temperature. This results in the water temperature at the outlet being lower than the temperature desired until this cooled water is first used. Obviously, the longer the run from the water heater to the outlet, the longer the waiting period for the water in the pipes that has cooled to room temperature to drain out or be used up.

In selecting a location for the water heater, a point should be chosen that is as close as possible to the point of use. Where more than one outlet is supplied by the heater, it should be placed nearest the point of most frequent use. For average residential use, this point will usually be the kitchen. However, for maximum efficiency, the heater should be equally divided between the kitchen, baths, and laundry areas.

Where an extra-long pipe run is encountered to a high-use area, it might be best to use more than one heater. For example, if the kitchen and laundry areas are close to each other, the main water heater should be placed close to these areas. However, if a master bath is located on the opposite end of the house, depending on the amount of use, it may be best to install a small water heater near the master bath for supplying hot water to only the master bathroom.

The water heater should not be located so it will be exposed to low surrounding temperatures, and wherever possible it is desirable to place the heater away from outside walls. The basement is the most popular location for the domestic water heater. This is fine if the basement area is heated. However, if the basement is unheated, it is better to place the water heater in a different location, such as a laundry or utility room.

In selecting the size of pipe to use, the pipe should be as small as possible to fully supply the outlets on the line. Pipe of a larger size wastes heat, because it contains a larger mass of metal to be heated during the use of hot water, affording a larger radiation surface. Furthermore, larger pipes will hold larger quantities of water that will cool between use at the outlets.

Pipe is sized according to the gallons per minute required at the outlet. For example, the average sink and lavatory faucets will draw approximately 3 gallons of water per minute, the average shower about 6 gallons per minute. Automatic dishwashers and clothes washers require large quantities of hot water, but usually only for short periods of time. Separate pipes should be properly sized and installed for these appliances.

The pipe material also has a bearing on the size of pipe to use. In almost all cases, copper pipe is the type currently being used. It requires the least amount of heat to bring it up to a given temperature, and copper pipe lasts almost indefinitely. On the other hand, galvanized iron pipe is heavy, requires more time to install, and will eventually rust out.

The amount of heat lost due to radiation from the water lines is dependent on several factors, but it can be reduced in all cases by some type of insulating material. All exposed hot-water lines should be insulated with an approved type of insulation.

In areas where hard water is prevalent, containing a large amount of soluble calcium and magnesium salts, the use of a water softener in conjunction with the domestic water system is recommended, as the deposit of scale on the heating elements will greatly reduce the efficiency of the elements.

Relief Valves

Water cannot be compressed, and it expands when heated. Therefore, some safety precautions must be supplied to ensure that the system will be safe to use. The pressure relief valve is the most common device to protect a hot-water system.

The most common pressure relief valve consists of a disc that is held closed by a spring. This type of pressure relief valve is installed in the cold-water inlet as near to the heater as possible. Whenever the internal tank pressure reaches the point to which the valve was set, the spring tension is overcome and the valve opens. The excess water flows from the valve and the tank pressure is reduced. See Figure 12-10.

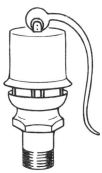

Figure 12-10: Typical water heater relief valve.

Combination pressure-temperature relief valves are also widely used. These valves are always mounted at the top of the tank where the water is hottest. They usually include a thermostatic element that extends into the tank to sense the water temperature. In use, when either the internal water pressure or temperature exceeds the setting, the valve will open.

Tank Protection

Most modern water-heater tanks are glass lined for protection. However, some of the older types utilized a magnesium rod installed in the water tank to reduce the rate of corrosion. The rod is normally installed through the top of the tank and extends very close to the bottom. The various chemicals in the water then attack the magnesium rod instead of the water-heater tank. When the rod becomes corroded to the point of being of little use, it may be easily replaced. Rods are normally provided in sections for ease of installation. If they are not in sections, it may be necessary to tilt the water heater in order to insert the rod, depending on the ceiling height above the water heater.

INDIRECT WATER HEATING

Thus far, only water heaters that heat water by direct application of heat have been covered. The indirect method, however, is often preferred when the main heating system is of the hot water or steam type. The indirect system permits furnishing year-round hot water in ample quantities at a reasonable cost via indirect water heaters that are adaptable to either steam, vapor, or hot water heating boilers of any size or type.

Indirect heaters consist of an arrangement of coiled or straight copper tubes for transferring heat from boiler water to the domestic water system. Two types are normally used: internal submerged and external submerged. The internal type is screwed into or bolted onto the boiler, while the external type is self-contained; that is, the copper tubes are encased in a steel or cast-iron jacket.

When external type heaters are connected with suitable piping to the boiler, hot water circulates through the heater casing, transferring its heat to the domestic water flowing through the copper tubing. Boiler water and domestic water are, of course, kept separate and never mix.

Indirect water heaters will supply hot water in all seasons whether the boiler is needed for space heating or not. However, it is necessary to install a suitable hot-water control below the water line to avoid getting the boiler temperatures above the steaming point during times when heat is not required in the rest of the home. This control, when set correctly, and when used with a properly sized heater, will provide abundant hot water at all times.

On a forced-circulation hot-water boiler, the temperature of the boiler water is similarly kept within certain limits. Circulation to the radiators is prevented by a flow-control valve installed in the supply main, which remains closed except when the room thermostats call for heat and when the circulating pump is running, or when the boiler is inadvertently overfired — although if properly controlled this latter instance rarely happens.

Even some older homes with gravity hot-water systems can have the benefits of indirect water heating provided a motorized valve is installed in the supply main to prevent circulation to radiators during warm weather. Such valves are electrically controlled to open only when the room thermostat calls for heat. However, in most cases, gravity systems have been updated to 100% automatic operation by installing circulating pumps and electronically controlled valves.

With indirect systems, the heating boiler is, of course, in constant use during cold weather, and only a small fraction of the heat generated is used in heating domestic water.

In summer or warm weather, the high efficiency of the boiler allows relatively economic operation, because once the furnace or boiler is brought up to the proper temperature for domestic water heating, it can be maintained at that degree on very little fuel. A few operations daily of the automatic firing device are usually sufficient.

Continuous use of the heating boiler during the times it is in operation for heating domestic hot water is also very good for the system. It prevents costly deterioration, which occurs when the boiler is idle for long periods. The internal combustion of the boiler is also improved by the use of an indirect heater, increasing efficiency and reducing fuel cost.

Indirect water heaters are available for use with or without storage tanks. Storage tanks require fewer square feet of copper heating surface to accomplish any given water-heating job, because hot water is stored in the tank against periods of peak demand. Such tanks are available in submerged, built-in, and external models.

The tankless heater, on the other hand, must have a large heating surface area so that it can provide hot water instantly and in sufficient volume to handle peak load demands. The tubing must be of the proper length, and the incoming cold water must pass through every lineal foot of it. Several factors must be taken into consideration when selecting a tankless heater. Some are pressure drop, volume of water, and boiler tax. Tankless heaters are available for installation within the boiler itself or for external application.

INSPECTING THE PLUMBING

The plumbing system's ability to furnish domestic hot and cold water, and to carry away waste is very important to the overall usefulness and value of any home. The following items are common problems that occur with residential plumbing systems; they are listed in their order of importance:

- Improperly installed drain traps and/or vent pipes.

- Loose joints in either supply and waste pipes.

- Rusted or corroded pipes.

- Concealed leaking pipes.

- Exposed leaking pipes.

- Uninsulated cold-water pipes.

- Leaking or dripping faucets.

- Broken or inadequate plumbing fixtures.

- Lack of air chambers in hot and cold water supply pipes.

- Pipe sized incorrectly.

Plumbing Systems

- Insufficient hot water.

- Deteriorated water seals.

Improperly installed drain traps and/or vent pipes can allow trap water to be siphoned out, leaving a clear channel for dangerous sewer gas to come back into the house. One way to determine if the drain trap and vent system is working properly is to fill the sink or basin with water, and then drain it. Listen for sucking or gurgling noises after a fixture is drained. Such sounds frequently indicate that the drain system is not vented correctly. If you detect foul odors around any of the waste drains, it's most likely sewer gas, and this is one gas you don't want to mess with. While not as quick as cyanide, it can kill just as deadly. If such a condition is expected, you should notify the owners or your client, and have them hire a plumber to check the system as soon as possible.

Loose joints are usually difficult to detect in most homes; simply because most of the plumbing pipes are hidden between walls or under floors. However, there are ways to "x-ray" these areas if you know the tell-tale signs. First of all, however, look for any exposed areas in the home; that is, areas without wall, floor or ceiling finishes. The basement in most homes offers these conditions. The attic space may be another area that exposes plumbing pipes. When exposed plumbing pipes are found, check these over carefully. Chances are if the pipes in these areas are properly installed, with tight, non-drip joints, the rest of the plumbing throughout the home will be in the same condition. To double check this theory, look for water stains in ceilings or walls — especially in a basement recreation or laundry room — where a finished ceiling has been installed. If water stains, or other evidence of dripping pipes are detected, there is more than likely leaky joints, or else the cold-water pipes above have not been properly insulated. And all cold-water pipes will form condensation and drip if not insulated. If no evidence is found, the joints apparently are sound.

While checking these exposed areas, note the type of pipe. Plumbing problems are common in older homes with iron or steel pipes more than 40 or 45 years old. Copper, brass, or bronze pipes and related fittings were not used until the 1940s. However, galvanized steel and black iron pipes were still quite common into the 1950s.

In recent years, PVC (plastic) pipe is used almost exclusively for drain and vent pipes, and is currently being used for both cold- and hot-water piping to some extent. Still, copper is the most popular and will outlast iron or steel pipe by a wide margin. If there is any question about the type of pipe used, a magnet may be used to test the pipe material; that is, iron or steel will attract the magnet, but copper, bronze, brass, or plastic will not.

Rust or corroded pipes should be obvious if they are visible. Another method is to observe the appearance of the water when the spigots are initially turned on. If discolored water appears, there is rust in either the pipes, fittings, or the pressure tank.

Leaking or dripping faucets are readily detected by turning the faucet off, and then looking for the drip. However, sometimes the drip may be slow and take a while. To detect these slow drips, turn all faucets off, take a paper towel and dry the sink, and then tend to the other phases of your inspection. Later, check each basin or sink for and water that may have dripped since you dried each with the paper towels. Don't forget to check the shut-off valves at each plumbing fixture. These areas are also potential trouble spots. Dripping faucets and valves are not only annoying, but can also increase the water bill and waste energy (electric or gas) used for water heating, if the condition is not checked. Should any drips be found, the washers or valves need replacing.

Broken plumbing fixtures are usually readily detected. Look for chips in the porcelain, especially on water-closet tank tops, edges of sinks, and similar locations. Ones that have cracks should be replaced. While checking for broken or cracked plumbing fixtures, you will also want to note if adequately-sized medicine cabinets are installed in all bathrooms. There should also be storage for toilet paper, tissues and related items. Linen closets or at least a small storage space for towels should be available in or near each bathroom. Also inspect the towel bars and toilet paper holders in these areas.

Checking for correct pipe sizes is directly related to the water pressure and quantity in the home. Turn on all faucets in kitchen, bathrooms and laundry area. With all faucets turned on, flush the toilets — one by one — as quickly as possible. If everything is operating properly, there should be a strong flow of water at each faucet. If water trickles out any of the faucets, problems exist: rust or corrosion in the water lines, undersized pipe, and similar maladies.

Plumbing Systems

If the pipes make noisy, vibrating or knocking sounds when turned on, problems exist:

- No air chambers in hot and cold water lines.

- Loose fittings.

- Water pressure too high for system.

Each toilet area should be checked thoroughly. Flush each water closet to see if it functions properly. If it drains slowly, there may be blockage in the waste line. Also check the floor area around each water closet. If any water damage is found, there is probably a faulty water seal around the base of the water closet.

Check the water heater for signs of leaks and rust. Modern water heaters will last 20 years or more, although the heating elements should be replaced about every 10 years or so.

Finally, determine the type of water supply and waste-disposal system that is utilized in each house that you inspect. In towns and cities, both the water supply and waste are usually a part of the city municipality. In rural areas, the water supply is usually from a drilled well, and the waste-disposal system is in the form of a septic system, as discussed previously.

The inspection checklist in Figure 12-11 will help you evaluate the condition of the home's plumbing system. However, if any doubts exist concerning the safety of any plumbing system, recommend that an expert go over the entire system thoroughly.

INSPECTION CHECKLIST — PLUMBING

Plumbing Information; circle appropriate items

Water Main:		City	Well	Cistern
Sewage System:		City	Septic Tank	Other
Size of Water Main:		1"	1½"	2"
Number of Bathrooms:		1	2	More
Plumbing Pipe	Copper	Galvanized	PVC	Other

		Yes	No
1.	Are all exposed pipes free of leaks?		
2.	Are cold-water pipes insulated?		
3.	Are drain lines clean and vented?		
4.	Do you detect any foul odors around any drains?		
5.	Is there sufficient water pressure at each faucet?		
6.	Are there any signs of sewage backup?		
7.	Do all plumbing fixtures have shutoff valves?		
8.	Are any plumbing fixtures chipped or broken?		
9.	Do all water closets flush properly?		
10.	Is the floor around water closets free of water damage?		
11.	Is the water clear (no rusty or discolored water)?		
12.	Is there sufficient hot water?		
13.	Do any ceilings or walls show signs of water drainage?		

Additional features:_____

Physical and functional inadequacies:_____

Repairs needed:_____

Quality of overall plumbing installation:

Figure 12-11: Plumbing inspection checklist.

Chapter 13
Operating Techniques

In this chapter, we are going to look at the home inspection business. We will discuss the operation of the business, what special skills are needed, the office and field equipment, and most importantly, how you can succeed in the home inspection business. The need for home inspectors increases every year. Forecasts predict that by 1999 over 80% of the lending institutions throughout the United States will require the services of home inspectors. Think about this for a minute. This means that every time a home is purchased — either new or an older one — the home must be inspected before the loan and mortgage can be finalized. With the thousands of homes that change hands every week, you should realize that the sources for a good income are almost endless.

The home inspection business, for the most part, consists of thoroughly inspecting homes for lending institutions that are considering financing them for prospective buyers. The mortgage holders must be assured that their investment is protected. This is where you come in. As a qualified home inspector, you should be able to evaluate any type of home that you may encounter, and then give your clients (the lending institutions) an accurate, easy-to-understand report.

There are also many private lenders who will require your services for the same reasons . . . to protect their investment.

Individuals who are buying an older home will also require your services. Chances are, most of them will be spending their life savings for the down payment on a home. Once they move in, they don't want to be surprised with many major repairs. For example, let's say the home's interior has been recently painted, the floors refinished, and the outside trim nicely touched up. The home looks good to the potential buyers. So they make, say, a $15,000 down payment, move in the home, and feel content that they are secure in a new home of their own. Then winter comes. They turn up the thermostat but find that no heat comes out of the forced-air outlets. They call a heating service and then find that the furnace is "shot," requiring a new one to be installed — costing a couple thousand dollars.

About the same time, these same homeowners find little white wings on the basement floor. Some probing with an ice pick indicates that some of the timber is damaged. So a termite exterminator is called in, a carpenter is hired to replace the defective structure, and another $5,000 is spent. You should now begin to see how your expertise can be invaluable to all real estate transactions. You can help protect all concerned by assuring them the house is what it appears to be . . . and if not, you have the ability to tell them what needs to be done to put the house in first-class condition before a loan is made, or before the home is purchased.

Although this chapter discusses what it takes to run a home inspection business successfully, we will not tell you whether or not you should go into business immediately. Perhaps you should start a part-time business first to see how things go, before giving up your regular job. You might be surprised at the amount of money you can make in the home inspection business after your normal work hours and on weekends. If zoning laws permit it, you can start out in your own home; if not, a small office space can be rented. Subscribe to an answering service, or merely purchase a telephone answering machine. When you finish for the day at your regular job, drop by your office, check the messages and make the necessary contacts to arrange to inspect the houses at your convenience. A few hours a week can bring in several hundred dollars extra. Or you may decide to try the home inspection business on a full-time basis, increasing your chances of more work and more income.

Whatever your decision, the following sections cover some very useful information that will help increase your chances of success.

CHECKLISTS FOR GOING INTO BUSINESS

Are you thinking of starting a home inspection business? Do you have the urge to be your own boss? Before you try it, let's be sure you know what it takes to own and manage your own business. You should be sure that you have what it takes to be that person. As we will discuss, being the owner of a successful home-inspection business takes more than just the knowledge of building construction techniques and your ability to spot defects.

Starting a small business of any kind is always risky. But your chances of making it go will be better if you understand the problems you will face. In this chapter, we want to help you work out as many of them as you can before you start your own home-inspection business.

A Checklist For Yourself

The question of your success in the home inspection business often boils down to whether or not you are a self-starter who can use your time wisely.

Listed below are ten important questions for you to answer.

1. Are you a self-starter?
2. How do you feel about other people?
3. Can you lead others?
4. Can you take responsibility?
5. How good an organizer are you?
6. How good a worker are you?
7. Can you make decisions?
8. Can people trust what you say?
9. Can you stick with it?
10. How good is your health?

If most of your answers are favorable, you probably have what it takes to run a business. If most of your checks are between favorable and unfavorable, you might have trouble with some aspects of running your own business and improvement on your personal habits is in order before you start. In this case, you might want to consider a partner — one who is strong in the areas where you are weak. If most of your answers unfavorable, your best opportunity for success probably is working for another home inspection firm.

More Considerations

Let's say that you have answered the ten questions above. Your answers indicate that you probably have what it takes to run a home-inspection business. Now, let's get into some more areas that you must consider before starting out on your own.

How about YOU? Are you the kind of person who can get a business started and make it go? There are around 500,000 new businesses started each year, but only about 40% of these make it. The remaining 60% close their doors within two years — often owing a lot of money to creditors in the process. A certain amount of money and enthusiasm will enable you to open your own home-inspection business. But it takes a lot of hard work, and many hours of your time to keep that business going and to pay yourself a decent salary. You'll find that you will have to do many things besides just inspecting a few houses each week. First of all, you have to make contacts with potential clients; negotiate a reasonable fee for your services; research the areas in which the homes are located to determine if the house you are inspecting meets with all local building and zoning codes; writing reports; billing for your services; collecting the bills; making (and meeting) operating budgets, including paying your own bills for utilities and rent on your office space; publicity and advertising so your clients will know you are available, and a host of other extra activities.

With the above in mind, think about why you want to start your own business, and what you have to get out of it. Are you retired or semi-retired, have office space available, and many of the required tools and office equipment. If so, you may expect a small income to supplement your retirement check and Social Security.

Do you have young children to support, a home mortgage, and car payments? Then you need a larger income. Perhaps you can start your home-inspection business on a part-time basis — working at your regular job in the meantime. Then, as you gain experience and get more and more clients, the day will come when you can make more money working full-time in your own home-inspection business, and you have risked very little getting there; that is, you maintained your normal income by keeping your regular job, yet were able to gain experience and make extra money working as a home inspector in your spare time. Sure, you may have to miss those Sunday football games at first, and you may not be able to take your family out to dinner quite as often as before, but you are building a foundation for a bright future. Many believe the sacrifice is worth it.

A big point that can affect your success or failure is "what is your regular job?" Have you worked in a business related to the one you want to start? Perhaps you have been a construction worker before. This is a good, solid foundation for the home inspection business. By the same token, any type of work in real estate or home appraising is another "plus" in your favor.

Have you had any business training in school? When bankers make startup loans to new businesses they always put any business training or previous business experience ahead of any technical training. You might be the most knowledgeable person around when it comes to building construction and home inspection, but if you are not a good business person, your chances of success is minimized.

Being self-employed means different things to different people. To you, as the owner of a home-inspection business, it means at least the opportunity to be your own boss, and to make money. But it really means more than that. As the owner, you must shoulder responsibilities to your community, to your clients, and to yourself.

The home inspector does have one very strong point in his or her favor. There is very little equipment to buy, and practically no parts or materials. You may have to replace a radon-testing kit every time you use one, but that's about it. Of course, you will have rent to pay, along with utility bills. There will be gasoline for your car to drive to the job sites. Furthermore, you will have some other incidental expenses, but compared to many other businesses, your cash outlay to start in the home-inspection business is minimal.

You are going to need some money to start, and it should not be borrowed money. How much have you saved? Are your personal finances in order? If you make your decision to start your business on a full-time basis, will you have enough money to live on while the business gets going? Don't be like the local fellow who decided to go into a certain business (it wasn't the home inspecting business). He spent over $30,000 on equipment, quit his regular job, and sat in his new office waiting for the work (and money) to pour in. A year later he was still sitting in the same place with not one job during all this time. His community had no need for the services he had to offer.

Therefore, it is strongly recommended that you line up some clients before making a big jump. Take a few days off from your regular job and investigate the situation. Talk to local bankers and others who may need your services. If they tell you that a dozen or so home inspectors come in on a daily basis looking for jobs, you might want to consider a different location. On the other hand, if these same bankers tell you they have been trying to find a reliable home inspector for their bank, and the ones they did find were booked up for three months, then that's your cue! Get a commitment to do their work, and do it right. If you put forth the effort and utilize what you know, you have a grand opportunity awaiting you.

To illustrate the importance of making sure there is a need for your services, let's look at a small Virginia town of a few decades ago. At that time, the town's population was approximately 7500 people. There were four gasoline service stations in town, all doing a booming business and the owners (and their employees) all making a good living. At that time, service stations washed car windows, checked the air pressure in tires, changed the oil when necessary, and even did some repairs.

The next year this number of service stations jumped to 32! A gas-price "war" started, service got worse, all of them got into financial difficulty, and eventually most of them went out of business.

So, make sure there is a need for your services before opening up a business next door to another home inspector. Competition is good for almost any business, and the need for home inspectors is high at the moment. Consequently, you should not run into a highly-competitive situation. If you do, however, think about opening your business in another area — perhaps in the next town (20 or so miles away) where there is more need for your services. A 30-minute drive from home, for success, is better than trying to compete with a handful of other inspectors who may already be in business across the street.

Starting Costs

To give you an idea of how much money you will need to start a home-inspection business, let's assume that you have no tools, office equipment, or other necessary supplies other than a car and a telephone installed in your home. The list in Figure 13-1 consists of the minimum equipment that you can get by with. Those items that you already have can be deducted from the total which should give you a good idea of how much you need for the basic tools and equipment to get started. Most of these items can be priced and obtained at any local hardware store. This should be the list to aim for as you work towards your goal, adding the items as you need them. Remember, these tools are a business expense and should be treated as such.

The only other item that you need is a supply of inspection checklists. There are several such forms scattered throughout this book. These can be copied and used in your actual inspections. Or, you can obtain preprinted forms. One source of these forms is HomeTech Systems, Inc., 5161 River Road, Bethesda, MD 20816. They produce very impressive forms at reasonable rates.

Living Expenses

The purpose of going into business is to make money. Have you figured out what net income per year you will need to survive? Then estimate the salary that you would like to make. Then take steps to make this amount, and at the same time, have as much of this guaranteed as possible. Let's explain that word "guarantee" a little more.

General Items

Ballpoint Pens
Business Cards
Detailed Road
Maps of Area.

Required Tools

Flashlight (extra batteries and bulbs)
Phillips-head Screwdriver
Flat-head Screwdriver
Circuit Tester
Inspection Mirror
Dial Thermometer
Ladder (minimum 12 ft.)
Binoculars
Hand Towel
Pocket Knife
Coveralls
Folding Rule
Ruler

Drawing Tools
Calculator

Helpful Tools

Small Level
Volt/Ohm/Ammeter
Camera with Flash
Steel Measuring Tape
Tape Recorder
Moisture Meter
GFCI Tester

Figure 13-1: Tools for the home inspector.

The only things really guaranteed are death and taxes; everything else is a "maybe" situation. But there are ways to make the odds more in your favor. For example, if you calculate that you must make a minimum of, say, $28,000 per year to survive; that is, the minimum amount of money necessary to make your current monthly installment payments, mortgage payment, utilities, food, clothing, and similar necessary living expenses, then you should have some means to make certain your income comes to this amount. Perhaps your spouse is working and making this amount already. While you were working at your regular job, this extra income heretofore has been for luxuries; that is, to pay for a beach house for two weeks each year; surf and turf once a week at a fancy (and expensive) restaurant; and an extra car or two. When you start into business, this "extra" money may now have to be used for necessities. That vacation may have to be postponed this year; surf-and-turf may take the form of dried beans and hamburger.

How Much Can You Expect To Make?

The home-inspection market will vary, but it will be in direct proportion to the real estate market. When the real estate market is booming, so will the home inspection business. The more homes sold and the more property titles that are transferred, the more your services will be needed. But the home inspector will also have work when real estate sales are low; even when a recession hits. Homes are constantly being put "on the block" for auction. This is necessary when a mortgage is in default. The bank must recover as much of the money as possible to satisfy its stock holders and to meet Federal regulations. Before doing so, however, the bank must establish a minimum price that they will accept at auction. To determine this price, the condition of the home must first be determined. This is where you come in. You will inspect the property thoroughly and report to the bank the overall condition of the house. With your report in hand, a home appraising team will then estimate the house's current market value; its estimated value in three to five years. From these reports, the bankers can estimate the home's investment value. If the house does not bring this amount at auction, the bankers will probably bid on the property themselves and try to sell it through the proper channels on their own.

Most businesses start out small and grow as the owner gains more experience, and as the business generates the profits to finance the growth. Suppose that the home inspection business in your area is currently limited, and those who are now in business cannot keep up with the workload. This would be a good time for you to start out on a part-time basis, say, inspecting one house per week for local lending institutions. At $200 per job, this totals about $10,000 extra income per year.

In working one day per week as a home inspector, you are gaining experience and making extra income at the same time. Yet, you still have the security of your regular job. Once you get the "bugs" ironed out and adjust your reports to the bankers' specifications, you become more efficient in your work — and the bankers are beginning to realize this. They know you are doing good work. If so, the time will come, when your clients will ask you to do an additional home one week. This means two homes instead of one on your day off. Eventually, this will turn into a regular thing; that is, you will have two homes every week. This obviously cuts down on your spare time, but you will then be making an additional $20,000 (at $200 per home) each year. Again, keeping your regular job.

If you continue to do work, charge a fair price, and give good service, more work will come in until it may be wise to consider going into the home-inspection business on a full-time basis. When you do, and if the work is available, you can expect a yearly salary of not more than $52,000 per year. This is about all the work one person can do; again, at $200 per job.

However, if the workload continues to increase, you may want to consider hiring other inspectors to work for you. Some people are darn good workers, but are just not business oriented. Some of these may live nearby. They have tried the home-inspection business on their own and although they did good work they never seemed to get things as organized as they should, and the business is failing. This type of person is an ideal candidate for an employee with your firm. In other words, you do the organizing, obtain the clients and handle the money while this other person helps you handle inspection jobs. There are all sorts of ways to go about this, but one way is to charge $200 for the job and pay your employee half of this plus any expenses incurred. You keep the rest in your pocket. This is your part of the profit for obtaining the work and organizing the business — seeing that everything is done right. You still inspect the same number of homes yourself, but you are gaining extra income from hiring an additional inspector to inspect even more homes.

If the workload continues to increase, up to three more employees can be hired, but four people is about all that you can manage yourself. Get above this number and you'll have to hire an office manager. Even with four people, you won't be able to inspect as many houses yourself because you will have to use some of your time organizing the work for your employees.

How About A Partner?

In opening your own home-inspection business, you have probably thought about going it alone, or perhaps having your spouse or other members of the family answer the phone and keep records. Today, this is not always a totally satisfactory arrangement. In starting a business, you most likely will need some money and know-how that you don't have. Possibly your relatives don't have exactly what the business needs, either. One solution is for you to find someone who does have exactly what the business needs. Then, that person must be willing to join with you in sharing the work and the profits and losses. This is the role of a partner.

When considering a partner, be sure that your prospect is a person with whom you share mutual trust and respect. Most importantly, your partner must be someone with whom you can get along, and without whom, your business may not get started. When bringing in a partner, or hiring employees, keep one question in mind. Each of these people expects money from your business. How many people can your business support?

Generally, a business may be organized in any one of three different ways. It can be a sole proprietorship, which means that you and you alone are the owner. Or the business may be organized as a partnership, as discussed above. In a partnership, you and one or more other people join together to own the business and to share in the profits or losses.

Should the business get large enough, you might consider changing its form to that of a corporation. This means that a group of people buy shares of stock in the business with their money. They are issued stock certificates in return for their investments. The stockholders own the corporation, and all they can lose is the amount of money that each invested. Traditionally, however, the home-inspection business does not lend itself to selling shares of stock. You may want to incorporate to protect your personal property against law suits and the loss of

Operating Techniques

personal assets should the business fail, but seldom will you sell shares to stockholders.

There are advantages and disadvantages to each form of business. For example, when you first start a business, being the sole owner is probably the best. It requires the least record keeping. While the amount of money you make is relatively small, you are required to pay less tax on it. But should you take on lots of inspection jobs, hiring many employees to help you with the work, your total gross volume could be thousands of dollars each week. Then, you might be advised to set up a corporation. It would pay less tax than an individually owned business doing the same dollar volume of business.

Important business decisions should not always be made by you alone. There are too many things that a person going into business has to consider. Some of these things you may not even know about, yet there may be legal penalties for doing them, or not doing them.

While considering starting your own business, talk with your lawyer about it. There is nothing wrong with seeking the advice of your family and friends on matters they know about. But, there is always the possibility that they will tell you what they think you want to hear.

Your lawyer, your banker, and your accountant are professional people. They know about the laws and other rules and regulations that may affect your business operations. It is their job to tell you the way it is, as they see it from their standpoint. Sometimes you may feel that you have paid your money to buy professional advice that you don't want to hear. But if you don't like what you are told, weigh the advice very carefully. These professional people will make more money from you in the future if you go into business and are successful, than if you fail. Also remember that as a home inspector, your clients may not always like what your report says. It might contain a whole list of defects with the home you are inspecting. However, you would never turn in a false report just to satisfy your clients. Neither will most lawyers, bankers, and accountants.

While the decision to start a business is yours, just as the day-to-day operating decisions are yours, always consider what you are being told and who is telling you. If the last home-inspection business that opened in your area closed when real estate sales dropped and never reopened, what will you do differently? For most people, a year-round income is needed to meet year-round expenses.

GETTING STARTED...RIGHT!

Why are some businesses more profitable than others of the same type and size? The answer is that the owners of the more profitable businesses keep their operations pointed in the right direction. They never lose sight of the goal of a business — to make a profit.

In this section, we make suggestions that should help you zero-in on profit. The successful owner of a profitable home-inspection business must keep informed, make timely decisions, and take effective action. To do these things, the owner must control the activities of the business at all times rather than being controlled by them.

People continually learn new things, and the home inspector is no exception. Consult building contractors, appraisers, realtors and building code officials to determine what kind of market is out there. Next, start a library. You will need technical building references as well as copies of the local building code and zoning ordinances. The local building inspector and you have very similar jobs. Go to his office and inquire about the publications that he uses for reference. Become familiar with magazine publications dealing with housing and construction. You may want to subscribe to a few of them (these will be legitimate business deductions on your income tax returns).

Your Office

Not only must an inspector set up organized and efficient methods for job site inspection but he should also set up his business office in a professional manner. A professional office not only includes good bookkeeping practices but also a good reference library and an organized storage system.

You should already be acquainted with building construction so the business of inspection probably seems like a great way to take your knowledge and put it to work for profit. Great! But don't ever take on the attitude that you know it all. The building construction industry is constantly changing. Things that are modern today will be obsolete tomorrow, so take steps to always keep abreast of the changes.

If the zoning laws in your area permit, an office in your home has many advantages. First of all, you won't have additional rent to pay, nor will your utility bills (other than perhaps your telephone) be much greater than they are now. You will be keeping initial expenses to the bare minimum which will help to insure your success as a home inspector.

Some people, however, are not motivated to work out of their home. They cannot adjust themselves to "working" when in their home. For example, you may have promised four inspection reports to your clients before tomorrow morning, and it's now 6 pm. You have plenty of time, if you work at it, but a certain television program may take your attention instead. When the tv program is over, you're too tired to work on the reports, so you go to bed. The next morning, when you only have two out of the four reports to turn in to your clients, they may think twice before giving you another job.

Working at home requires a certain amount of self-discipline, and if you don't have this trait, you must take steps to acquire it. If this seems impossible, then you should think about renting an office away from home, or else forget about going into business for yourself altogether.

But let's say you have what it takes. You are all set to set-up "shop" in your home. You also have your family's assurance that they will help in every way possible; that is, when you're working in your "office," your spouse must take the attitude that you're "away from home." If that garbage can needs emptying, she should wait until you're finished your work before asking you to carry it out to the curb for pickup in the morning.

Unlike many other types of business, you don't need an elaborate office setup to enter the home inspection business. Seldom will you have to greet the public in your own office. Most of your work may be categorized as "house calls." You will be "in the field" when you must meet the public.

However, you do need a place to work, preferably away from the normal family activities. This could be a spare bedroom, den or other seldom-used area in the home. If none are available at the moment, you might consider building an office area in your basement or in the garage. If these latter two routes are taken, you should figure on spending a couple thousand dollars on the remodeling, heat, air conditioning, and electrical work required; maybe more.

You should have a desk and chair, storage drawers, a telephone, and an answering machine. With just this much office equipment, you would be surprised at the amount of work a home inspector can accomplish. If you purchase new equipment, you can count on spending at least $500 for just these few items. However, many businesses close their doors on a daily basis; others change locations and must get rid of their old office furniture before moving. Many home inspectors have picked up a practically brand new desk and a couple of filing cabinets for less than $50 this way. It takes some looking around, but deals like this are plentiful.

If you purchase a full-size desk for your office, you probably won't need a filing cabinet for the first weeks, since most desks have a few filing drawers built in. Then after you get a few inspection jobs (and have been paid for them), you might want to consider purchasing a three-drawer filing cabinet to keep your client's files in order.

Your local telephone company will connect the phone for you, and an answering machine may be purchased at almost any department store or office supply company.

The next item that should probably go into your home office is a bookcase. As mentioned previously, you will want to collect a large library of reference manuals. Many of these can be obtained at no charge from manufacturers of building materials. Others will have to be purchased from the publishers. Therefore, you will want a rather large bookcase to start out, and you will probably find this inadequate in a short time as the size of your library increases.

As the number of inspection projects increases, you will have to add another filing cabinet or two, but don't purchase these until you actually need them, unless of course you find a deal whereas you can get them for a very reasonable price. Then buy three or four and keep the spare ones in a storage area in your basement or garage. And you will need them if you continue in the home inspection business. To be organized, you must have a separate file for each of your projects, filed in a manner so that you can find the file folder at a moment's notice. Some inspectors like to have a separate filing-cabinet drawer for each of their larger clients — like bankers and other lending institutions — and then a separate file in these drawers for each project inspected. Others maintain files in a chronological order; that is, by the date the job was inspected. Either way is acceptable; choose the filing method that suits you the best.

Office Layout

The home inspector's office will vary in size and layout, but as mentioned previously, a separate area should be provided for the purpose of doing your office work so as to be out of the way of normal household activities. The office area should be illuminated with 100 to 150 footcandles of well-diffused light, painted a pleasing color (preferably a pastel shade), provided with adequate heating, cooling, and ventilation, and of sufficient size to accommodate all necessary office furniture, filing cabinets, drawings, and the like.

The furnishings within the area should consist of a large desk, perhaps a utility or "throwoff table," comfortable seats, filing cabinets, and you might even consider a blueprint file because you will eventually have many sets to contend with. Bookshelves for manufacturers' catalogs and reference books, a magazine rack for trade journals and other construction magazines. All should be in easy reach of the home inspector while seated at his or her desk. If required, you may also want a small drawing board with drafting tools to "clean up" your floor plan sketches of homes that you made at the job site. The neater and more professional-looking your drawings are, the better impression your clients will get of your services.

Some home inspectors like to take measurements and make rough sketches on the job site, and then use a computer CAD system to redraw these to the highest quality. If you already are using a computer, many CAD software packages are available for around $100 that will enable you to turn out professional-looking drawings, even on an inexpensive dot matrix printer. This is something to consider. Remember, you may not be swamped with competition at the moment because the home inspection field is relatively new. However, as others see your success they will want to give it a try also. Do your best work and you will be one jump ahead of the competition.

The home inspector should be equipped with an electronic calculator — one that can be used in the office and also on the job. A small pocket calculator is ideal for both, although some inspectors take the pocket calculator to the job site in their briefcase, but use a larger console calculator in the office. The larger calculators are easier to read and the keys are bigger, making it easier to use. The larger calculator, however, can come at a later date, when you have a few jobs under your belt and your bank account is a little larger.

Office Activities

While much of your time will be spent in the "field" actually inspecting homes for your clients, much time will also be required in your office. Some activities performed in your office will include, but is not limited to, the following:

- Writing reports from your checklist information obtained at the job site.

- Billing clients.

- Paying business-related bills.

- Researching a project.

- Studying blueprints of your projects.

- Comparing the original blueprints with your on-site sketches to see how closely they match.

- Redrawing your sketches into professional-looking documents.

- Bookkeeping.

- Updating client files.

- Organizing your files so they will be readily accessible.

You should have the space and facilities to perform all of the above. If not, take means to acquire it — either now or later.

COST ACCOUNTING

One weakness of those entering the home-inspection business is their failure to recognize the importance and need of good accounting records and related bookkeeping procedures. Hip-pocket bookkeeping

Operating Techniques

systems are common in one-person operations, and this is one of the reasons why so many new businesses fail within the first year or so.

Every home inspector needs some system to enable him or her to bill clients for jobs performed (keep track of what money is received and what is still due), keep track of the amount due others, and maintain required records and other general operating expenses. Another important reason for bookkeeping systems is to provide a basis for preparing federal and state income tax reports.

The objective of this section is to emphasize the need and benefits of proper accounting records and to provide some guidelines to follow in selecting the more applicable bookkeeping system for each home inspector's individual needs.

Need For Good Records

The benefits of good accounting records for a home inspector are many, but they can be summarized as follows:

- Records show what has taken place in the past and form a basis for determining why. They give the home inspector a look at his or her past, both good and bad, and give him a chance to correct the past mistakes as well as indicating the direction for better profits and business opportunities.

- Good records enable the home inspector to compare his operations with industry statistics and trends.

- Good records present an orderly and complete picture of the firm's financial position which lets the home inspector know exactly where he stands.

- The correct use of good records facilitates the preparation of payroll and tax returns.

- Adequate records can also help in the collection of accounts when invoices are disputed. On the other hand, they can prevent the home inspector from having to pay bills that are not due.

Choice of System

An accounting system integrates various bookkeeping forms such as *Sales Journal, Purchase Journal, Cash Receipts Journal, Cash Disbursement Journal, Payroll Journal, Accounts Receivable Ledger, Accounts Payable Ledger,* and perhaps a *Job Cost Ledger,* into a functional operating system. The system provides for the orderly entry of the figures onto the accounting records and an orderly basis of securing from those records the needed or desired information and end products in the form of client invoices, job cost records, financial statements, and so on. Different bookkeeping or accounting systems vary considerably in detail and method although all follow the same basic principle. The system that will best fit your need will depend upon the type and volume of work performed as well as your financial capabilities.

The new home inspector can get along nicely with a simple manual or hand method of bookkeeping. Small systems of this type are available from a number of sources, ranging in price of about $15 to around $100. These systems are designed to enable those with little or no bookkeeping experience to analyze and interpret the progress of their business, including the hidden sources of profit and loss. These systems (which are available through most office suppliers and office supply catalogs) combine journals, ledgers, payroll records, and so on, all in one binder that adapts very well to the needs of the home inspector.

Although the manual bookkeeping system works well for the home inspector just starting out in business, after a certain volume of business is reached, a computerized accounting is usually the best route to take. Many people find these simple computer bookkeeping programs much easier to use than manual bookkeeping systems, and many suitable programs can be purchased for a price between $50 and $100.

The advantages of a computer accounting system over a manual system are as follows:

- A computer can do all types of arithmetic functions many times faster than a human can; and computers (when operating properly) are virtually error free.

- Computer systems can be programmed to do a great number of functions automatically. Thus instead of a bookkeeper having to remember certain things to do at a certain time, the computer does the job automatically.

- By using a printer and the right type of paper, the computer system can automatically print all of your checks, invoices, statements, memos, and so on.

The primary functions of a computerized accounting system are as follows:

- *Billing and Collections:* In addition to automatically printing and maintaining a record of every billing, the computer system can be set up to automatically send out late notices when accounts reach a certain time.

- *Payables:* As stated earlier, the computer can be used to automatically print checks. It will also maintain a permanent record of bills that you have paid. Payables can also be set up to be paid on a certain date or in a certain order.

- *Periodic Statements:* The computer can generate monthly, quarterly, and yearly statements automatically, thus showing your financial status at any given time. With the proper type of program this can also be broken down into different types of expenses and incomes, helping you to pinpoint profitable parts of your operations and also trouble spots. The statement can also be arranged to show the current year's figures against the previous year's figures, or any number of different ways. With the programs that are now available, the options are almost limitless.

- *Monitoring Cash Flow:* The computer can keep track of your bank balances, billings, payables, credit lines, and so on. By just pushing a few buttons, you can have all of your current cash figures shown together on your monitor for your review. This is very important in helping you to know exactly where you stand at any point in time.

PUBLIC RELATIONS

There are several factors that will help insure your success as a home inspector. Knowing the business is one of them; proper management is another. One of the most important, however, is public relations — selling yourself and your business.

Any small-business owner wants to project the best possible image of the business. Take a look at yourself in a mirror. You want your present and potential clients to get a good impression of your business and of you.

But what kind of image should you have? First, don't think of your work as being in any way inferior to that of a neighbor who may be a doctor or lawyer. As a qualified home inspector, you are also a professional — the same as a doctor, dentist, lawyer, engineer, and others. In fact, these people may soon invite you to join the local community service clubs they belong to: Chamber of Commerce, Lions Club, Rotary, and similar organizations. Shortly after you start in business, and if you do your part, your particular qualities and abilities will be recognized. Then you may be asked to carry a share of the community load in perhaps building a new hospital wing, putting the local charity fund drive over the top, or even running for public office.

Customer Relations

Your business depends on how well you serve your clientele. More and more of these people and organizations are becoming particularly conscious of the services you have to offer. Once you have proven yourself, they will listen carefully when you say, "the electric wiring in that last house I inspected is not only dangerous to this house, but also to those next door" or "I recommend you have this house treated for termites before making a loan on it."

Now how about problem clients? Some of them will eventually hire your services, and regardless of how hard you try or how good a job you do, they will never be pleased. In such cases, always be friendly and courteous. You may wish that they take their business elsewhere the next time, but be as pleasant and as courteous as you can. If you tell an unpleasant customer to take his business elsewhere, the customer or client may do just that. In the long run, this hurts you and benefits your competition. Many businesses have failed because of poor client relations. In most cases, however, the home inspector deals with professionals, like him or herself, who must also be pleasant and courteous in their jobs. Seldom will you run into a crank, needler, or malcontent when dealing with officers of lending institutions.

Community Relations

As a business owner, you will be accepted into the business life of your community. Accordingly, as with memberships in a club, you will have obligations. And they will cost you time and money. If you pay your dues at a club but never hit a golf ball or swim a stroke, that is your personal business. However, if you fail to capitalize on your position in the community, it may cost your business.

Rotarians like to do business with Rotarians. Parents of Boy and Girl Scouts like to buy from parents of other scouts. In general, people like to deal with other people who are active in community affairs, people who have shown they can assume responsibility with United Fund campaigns or YMCA building budgets.

Never be bashful about getting your picture in the newspaper as a member of a committee. People like to see you supporting community projects. They even expect it of you.

The dollar amount of your civic and church donations should be figured in your advertising and promotion budget. Many businesses donate a very small percentage of gross sales. If your advertising and promotion budget is stretched already, consider sponsoring a Little League baseball team or a bowling club. This may show your community involvement while conserving your cash.

The amount of time you can devote to community activities and club meetings depends largely on the amount of free time you have away from your own business. Don't pledge time to community projects and then have to back out because you have physically overextended yourself. "Moderation in all things" is a saying that certainly must be applied to your civic activities.

ADVERTISING

Traditionally, advertising among professionals has been a no-no. There is no law against professionals advertising; as a matter of fact, many attorneys and even some doctors are now advertising their services on television. Still, publicly advertising professional services is in poor taste; it should be beneath your dignity. Business cards and letterhead are certainly in order. There is also nothing wrong with personally visiting each of the lending institutions in your area and leaving a brochure or your business card with the mortgage loan officer. You may want to also pay all realtors in the area a similar visit.

You also want to make sure that your name is listed under the appropriate category in your local Yellow Pages. You may also consider listing your name in the Yellow Pages of adjacent communities.

Not only is radio, television, and newspaper advertising in poor taste for a professional such as yourself, it will do very little good in promoting your business because most of your clients will be the local bankers, realtors, or other lending institutions. A personal visit to each of these establishments will do far more good than an ad in the local newspaper.

However, your local newspaper may want to do a story on your new business. This is one way to let people know you are in business. In fact, most small-town newspapers give all new businesses the courtesy of a one-time story to give them a jump start.

You will also have a chance to let your services be known at the various civic functions. How many times have you heard someone ask, "what kind of work do you do?" Here's your chance! "I'm a professional home inspector."

Some people may want to know more details about this home inspection business, and you should be in a good position to tell them. One of them might happen to be the president of one of the local banks — the one that makes more mortgage loans than any others within 20 miles. If so, he may ask you to pay him a visit the next time you are in the vicinity to discuss some business. You are then well on your way to success as a professional home inspector.

Chapter 14
Preparing Reports

A *Report* is an account or statement describing in detail an event, situation, or the like, usually as the result of observation, inquiries, etc. Reports made by home inspectors, in general, deal with the type and condition of a residence — along with the related electrical and mechanical systems.

Although a report may be submitted in several different ways, the better ones use the fewest words, yet cover everything of importance in a concise, clear manner. One of the best ways to record information concerning your home inspections is to use a well-prepared, preprinted form especially designed for the home inspector. A home inspection form is a systematic method of presenting notes or the condition of a home in tabular form. When properly organized and thoroughly understood, home inspection forms are not only powerful timesaving devices for the home inspector, but they also save those who must read them much valuable time.

Regardless of the type of report used, all should have at least the following information:

- Client's name and address

- Complete address of home being inspected

- Description of home

- Description of neighborhood

- Evaluation of home (structure, finishes, grounds, building systems, etc.)

- Statement of conclusions

Besides this information, all reports must be legible; another reason that pre-printed forms can be helpful. When a space or box on the form is merely checked, the printed information beside the box will be highly legible. Should the lending institution or others require a different format, the information you obtained on your field report can be readily transferred to a different format on your computer.

This chapter is designed to show you what information to include in your home inspection reports, organizing various types of reports, compiling reports, and finally, how to submit them to those hiring your services. Both manual and computer-aided techniques are covered.

TYPES OF REPORTS

In general, there are three basic types of home inspection reports:

- Narrative

- Blanks

- Checklist

The type to choose will depend mainly on the home inspector's preference, or the requirements of the lending institution, but many home inspectors choose a combination of all three styles. That is, the home inspector will use a form with combination checklists and blanks for his or her on-site inspection. The data on this checklist/blank form are then used to compile a narrative report that will eventually be submitted to the lending institution hiring the inspector's services.

Narrative Reports

A narrative report is the type often recommended by many trade organizations, law firms, insurance underwriters, and lending institutions. However, it is usually the hardest type to write. Therefore, it is recommended that a checklist-type form first be used to collect data that will go into the narrative report. A sample narrative report appears in Figure 14-1 on the next page, and a brief description of each item is given in the following paragraphs.

Looking at the report in Figure 14-1, from top to bottom, note that the date of the report is placed directly above the client's name and address. This is to indicate the date the narrative report is written, not when the porperty is inspected, although both dates could be the same. Next comes the name of the property inspected, including the full address and date the inspection was made.

The first paragraph of the report gives a brief description of the house and property with an overview of the general condition.

The second paragraph lists a major defect that decreases the value of the house; that is, the paint on the southern side of the house is blistering and peeling. This should be corrected by scraping and repainting as soon as possible to prevent further deterioration of the siding.

Looking at the third paragraph, we find that this property has a major safety hazard. The concrete apron on an old cistern, which is no longer in use, is cracking and showing signs of a possible cave-in. By all means, the cistern should be filled in and capped before the property owner has a major lawsuit on his hands by someone falling in it.

The fourth paragraph is a miscellaneous one with the subject being "other items needing attention." A few loose bricks on the chimney cap need resetting, splash blocks need to be added at the downspouts, a rotten wooden screen door needs replacing, the rear entrance door needs to be refitted, and the attic needs more ventilation.

The next portion of the report deals with the major components of the house and the following should be noted:

1. Foundation is poured concrete with a crawl space
2. The siding is yellow pine clapboards (in need of a paint job).

July 27, 1995

Mr. John T. Adams
Vice President—Home Mortgage Loans
Jefferson National Bank
44 Broad Street
Overall, Virginia 22848

Re:
    Clatterbuck Property
    6677 Kaiser Avenue
    Rileyville, Virginia 22650
    Date of Inspection: July 26, 1995

This Dutch colonial frame dwelling was expertly constructed in the 1920's, shortly after a deed was issued in 1923 from Sheldon B. Davis to William J. Troeger, and recorded in the Page County Virginia Deed Office. The house is situated on approximately three acres of relatively flat land with good drainage. In general, the house is solid and in good condition, but may require some maintenance in the near future for the house to remain in this condition.

Major defects: Paint on southern outside siding and trim is blistering and peeling. Should be scraped and repainted in the near future.

Safety Hazards: Concrete apron on old, defunct cistern is cracking and could cave in. Since property now has a drilled well, cistern should be filled in and capped.

Other Items Needing Attention: Reset a few loose bricks at chimney cap. Add splash blocks at four downspout bases. Wooden screen door on back porch needs replacing (wood rot). Back entrance door (into kitchen) is warped; needs planing and refitting. Attic ventilation is inadequate; needs larger vents.

Figure 14-1: Sample narrative report for a home inspection job.

The major components of the house are briefly described as follows:

Foundation:	Poured concrete with crawl space
Siding:	Painted yellow pine clapboards
Roof:	Painted tin with rolled joints. Some rust spots showing through paint.
Heating:	Electric baseboard heat in all rooms except bath. Bath has electric down-flow wall heater.
Hot Water:	4500-watt electric water heater.
Electric Service:	200-ampere circuit-breaker panelboad.
Plumbing:	Copper water pipes; cast-iron drains and vent pipes.
Water:	Drilled well.
Waste Disposal:	Private on-site septic tank and drain fields.

Figure 14-1: Sample narrative report for a home inspection job. (Cont.)

Insects:	No visible signs of insects.
Utility Room:	On first floor with standpipe for washer and 240-volt dryer outlet.
Baths:	One full bath (tub/shower, water Closet, lavatory, and linen closet.
Basement:	None.
Interiors:	Lime plaster walls and ceilings. Wood paneling in utility room and den.
Attic:	10-inch fiberglass insulation. Ventilation inadequate.
Garage:	None

The above is an overview of the home's condition in general. A more detailed report appears on the pages that follow:

Sincerely,

J.P. Warren
Home Inspection Services

Figure 14-1: Sample narrative report for a home inspection job. (Cont.)

3. The roof is rolled-joint tin (which needs painting)
4. Heat source is all electric, the electric water heater is 4500 watts which should be adequate
5. The house has a 200-ampere circuit-breaker panelboard
6. The plumbing consists of copper water pipe and cast-iron drains and vent pipes
7. A drilled well is the water source
8. The property has its own septic tank and drain fields
9. There is no visible signs of roaches, termites or other insects
10. The house has no basement so the utility room is on the first floor and is equipped with a standpipe for washer; also a 240-volt dryer outlet
11. There is one full bath with a linen closet
12. The walls and ceilings are lime plastered with wood paneling in utility room and den
13. The attic is insulated with 10-inch fiberglass and has inadequate ventilation
14. There is no garage

It should be obvious from the format of the narrative report that an all-new report will have to be written for each project. Unlike other styles of reports (blanks and checklists), very little information from one report can be transferred to the next. There are too many variables to consider. Consequently, the report written in a narrative style will usually take longer than other types. Still, this is the type of report preferred by most lending institutions and the type also recommended by several home-inspection organizations. Therefore, learn to master this style of writing.

Blanks

In this type of report, shown in Figure 14-2 on the next page, the inspector fills in the blanks, after a brief walk-through inspection, with a word or words. This type of report isn't as satisfactory as other types because very little detailed information can be given. They don't convey many of the important details to the client and give little

	ROOF-ATTIC	
Item Number	**Description**	**Comments**
1.	Roof Type	
2.	Valleys/Flashing Skylights	
3.	Porch Roof	
4.	Overhangs	
5.	Attic	
6.	Roof Support	
7.	Insulation	
8.	Ventilation Condensation	
9.	Gutters and Downspouts	
10.	Plan View of Roof Structure	

Figure 14-2: A partial blank-type inspection form.

indication about how thoroughly the house was examined. For example, if the blank report has, say,

"Roof Type_____"

for its first "question" (and you fill in the answer in the blank space), you would have room on the report to indicate gable, gambrel, hip, etc., but if there were any modifications to the roof, or if it was not a typical roof, there isn't any room to go into detail about it. Therefore, you will perhaps have to leave out some detail that may be of interest to your client.

To illustrate this point, look at the roof type in Figure 14-3. Then look at the partial report in Figure 14-2. What would you put in the blank space provided for "House Roof" (roof type)? After studying

Preparing Reports

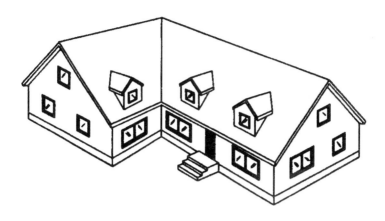

Figure 14-3: One type of house roof that the home inspector may encounter.

the view for a moment (Figure 14-3), we probably realize that we have two wings to the house, each of which has a gable roof. Since there is a *valley* where the two roof sections join, we could call this roof style *gable-and-valley roof*. But suppose one of these gable sections had two dormer windows on the front side and a shed (lean-to) dormer on the back. Your blank space would become a little crowded trying to get all of this information in the one blank provided on the form. It seems you would either have to leave out what might be important information or else use a longer form. But there is one more way: using the blank-report in conjunction with a photo. Put the words *gable-and-valley roof* on the form blank, and then take photos of the house from different sides. You may then print additional notes on the photos to more clearly show what style of roof you are really talking about. For example, "Note two dormer windows on north wing, each of which faces north; lean-to dormer on south side of north wing, facing south."

Figure 14-2 shows one type of blank report that may be used by the home inspector — either as his or her final report or else for the preliminary report . . . the latter being the recommendation. After completing this preliminary report, a narrative type report is them completed to be presented to the lending institution or other clients.

Checklists

A home-inspection checklist report has printed pages with blanks or squares to be checked for applicable variations. Additional spaces are usually provided for comments, remarks, and descriptions of items not included on the printed form. See Figure 14-4.

Most home-inspection checklists are elaborate affairs, with 10 to 20 printed pages. However, every page will not be applicable to each and every home-inspection job. Therefore, when the job has been completed, the home inspector will discard those pages of the checklist that do not apply, and then use the remaining pages as his or her report.

There are several advantages of the checklist over other types of home-inspection reports:

- Fast turnover. Report can be completed on the job site and given to the client immediately upon completion of the inspection.

- Ease of reading. A quick glance down the list will give an overview of the home's condition in a matter of minutes.

- Less writing requirements. The checklist report does not require as much of the inspector's time going over punctuation and forming good sentences to impress his or her clients.

- Less chance of omitting something. With a well-prepared checklist report, most items will already be printed on the form which will facilitate the inspector's job and prevent him or her from omitting anything from the list.

While there are several advantages of using checklist reports, there are also several disadvantages:

- Difficult to interpret. Due to the lack of detailed information, a checklist report may sometimes not be fully understood — requiring additional reports to be made.

STRUCTURAL

Type of Building	☐ Single ☐ Duplex ☐ Rowhouse/Townhouse ☐ Condo ☐ Other _____ ☐ Wood Frame ☐ Masonry ☐ Other _____ ☐ Gable Roof ☐ Shed ☐ Hip ☐ Gambrel ☐ Mansard ☐ Flat ☐ Other _____
Structure	Foundation Wall: ☐ Poured Concrete ☐ Block ☐ Brick and Block ☐ Solid Brick ☐ Other _____ Floor Framing: _____ Wall Framing: _____ Roof Framing: _____ ☐ Signs of Water or Insect damage ☐ Extensive ☐ None noted ☐ No major structural defects noted — in normal condition for its age

BASEMENT (OR LOWER LEVEL)

Basement	☐ Full ☐ None ☐ Open Walls ☐ Closed ☐ Open Ceiling ☐ Closed ☐ Extensive present basement storage, visibility limited
Floor	☐ Concrete ☐ Dirt ☐ Other _____ ☐ Satisfactory ☐ Resilient tile ☐ Carpeting ☐ Other _____ ☐ N/A
Floor Drain	☐ Satisfactory ☐ N/A
Sump Pump	☐ Operating ☐ Not Operating ☐ French Drain ☐ N/A
Basement Dampness	☐ Some Signs ☐ Extensive ☐ None Noted ☐ Past ☐ Present ☐ Not Known
Crawl Space	☐ Readily accessible ☐ Not readily accessible ☐ Satisfactory ☐ Conditions observed ☐ Method _____ ☐ N/A ☐ Conditions not observed Floor: ☐ Concrete ☐ Dirt ☐ Other _____ Clearance below joists: ☐ Ample ☐ Inadequate Dampness: ☐ Some Signs ☐ Extensive ☐ None Noted ☐ Vapor barrier ☐ Insulation ☐ Ventilation

Figure 14-4: One page of a checklist report.

- They are very difficult to amend should a mistake be made, or else the home inspector has second thoughts about a particular project.

- A handwritten report will not look as neat as a typewritten narrative report. If the inspector's handwriting is poor, this could cause problems.

Most of the chapters in this book have an inspection checklist covering the chapter subjects. Combining all of these checklists together forms the basis for a good checklist report.

Once you have gained some experience in the home-inspection field, you will no doubt come up with ideas of your own. Furthermore, experience will tell you which type of form suits you the best. In other cases, your clients may dictate the type of form they want submitted. Consequently, you should be prepared to submit any and all types of inspection forms.

The next section will cover details of filling out all types of inspection reports, from the time you get the job to the time you turn in your final report to your client.

PRACTICAL APPLICATIONS

Each home inspector has his or her own method of inspecting and reporting on homes for clients. To say that one method is better than another would be wrong. It is entirely up to the individual to choose a plan that suits his or her needs the best, and which will enable the best work to be performed. However, every new home inspector must have a foundation upon which to start. As more experience is gained, modifications to this basic plan will take place. Further honing and refining of this plan will eventually end up in a "perfect" one to suit the individual's own needs.

The following is how one home inspector performs his work — from obtaining the client to turning in the final report and collecting his money. After reading about this inspector's method, you may wish to design your schedule and techniques around this plan. But you don't have to. If a better method suits you, by all means use it.

Immediately upon being contacted by the inspector's client, the home inspector gathers all pertinent information concerning the project:

- Name and address of the client.

- Address of property to be inspected.

- Name and address of current owners of property.

- City and/or county in which the property is located.

- The amount to be charged for the inspection job.

- Any particular requirements from the inspector's clients.

- Date contact was made.

- Date of scheduled inspection.

- Date report is promised to client.

The home inspector also tries to find out exactly what type of home is involved; that is, the style of home, number of floors, type of neighborhood, age of building, etc. He further inquires as to available blueprints of the home, who built the house originally, how many owners have been involved since the home was built, and similar details. If blueprints of the home are available, the home inspector will obtain a set of these before commencing the inspection. He will also try to find out if the home has had a recent termite/pests inspection by a qualified inspector. In addition to this information, the asking price of the home may be helpful if the inspector is unfamiliar with the neighborhood. It takes more time to inspect a $500,000 home than a $70,000 townhouse.

Knowing what type of house style is involved will help the inspector select the equipment to take to the job site. While some inspectors use a small van or truck for their work — having sufficient equipment on board to handle every conceivable inspection job that could possibly be encountered — most home inspectors use the family car. Consequently, the equipment is loaded and unloaded for each job. If the home inspector knows that the house is, say, a single-story ranch style home, there is no point in loading up his 20-foot extension ladder on racks attached to the car's roof. Rather, his 10-foot ladder should be quite adequate.

If it's an older home, the inspector may want to take a lead-testing kit to see if any of the exposed paint contains lead. If a newer home, there is no need to take the testing kit. If the neighborhood is noted for radon gas problems, a radon check is in order. Otherwise, a radon-testing kit is not necessary, unless the inspector's client specifically requests it.

If the builder is known, the inspector may want to discuss the project with the builder. In fact, home inspectors should know about all of the major home builders in the area in which he or she inspects. Eventually, you will know which ones do the best work, and which ones might not do so good. Once you know this information, you can plan your inspection accordingly. For example, if past experience shows that one particular contractor always cuts corners in his building projects, providing structures that barely meet the code requirements, your client would appreciate knowing this information. It will also tell you that a more thorough inspection must be done on homes built by this particular builder.

Having a set of construction drawings for the property being constructed can save the home inspector much time. If available, the home inspector in question obtains a set of construction drawings (blueprints if you will), and briefly studies the floor plan and elevations of the project. Seldom will a finished house be constructed exactly like the blueprints; there are usually a few changes made as the project progresses. However, the home you are to inspect should be very close to the original drawings.

On many sets of construction drawings, you can also find out the type of heating equipment specified, the size and type of electric service involved, and a host of other important details about the project. These details can inform you as to the type of equipment needed for the job and it will facilitate taking measurements on the job site.

Since most lending institutions require at least a floor-plan sketch of the house in question, an already-prepared set of blueprints can make your work easier. You can take them to a graphic service and have copies made from the original sketch; this will save you the time of taking measurements on the job and then sketching out a floor plan yourself. You might have to make a few changes to the original set, but the majority of the work is already done for you.

So you're convinced; a set of blueprints should be obtained. Where do you get them? Remember, not all homes will have construction drawings available. In fact, most homes built prior to 1980 will probably not have blueprints available. However, for later-built homes, you should try the following:

- If the builder is known, contact the builder to see if an office set is still in the files.

- Since many lending institutions require floor plans for any construction loans, you can try the bank or savings & loan company that made the original loan.

- City and county tax offices normally have at least a sketch of the property, and many will have a set of complete blueprints.

- City or county inspection departments also often require a set of drawings to remain on file.

- If the house was designed by a local architect, this is the next place to contact.

- The present owners may have a set of construction drawings that were left with them after the house was constructed.

At The Job Site

We are now back with our original inspector, the one whose techniques we are now going to describe. Once all of the above information is obtained, this particular inspector loads the necessary equipment, notes the milage reading on his odometer (before leaving), and with detailed directions to the property, proceeds to the job site.

Upon arriving, his first pieces of equipment used (and carried with him) include: a clipboard with his checklist report attached, his camera (loaded with color polaroid film), and a small battery-operated hand-held tape recorder. A tape measure is attached to his belt. Armed with this equipment, he starts off at the top of his well organized checklist, checking the appropriate boxes; that is, type of home, roof style, siding, basement or crawl space, etc. If he comes to a category that is not adequately covered by the brief information on his checklist, a button is pressed on the hand-held tape recorder and he dictates additional information about that particular item. The recording may sound something like the following:

"House style is basically Cape Cod, but modifications have been added: two dormer windows on entrance side of house, on second floor; recent carport built on right-hand side of house (as I face the house from the front). Consequently, in my written report, the house style should be called *modified Cape Cod.*"

The inspector continued on down his checklist until every item has been covered. Conventional items are merely checked off on the list. Those requiring modifications are first checked off and then additional information is dictated on tape. Of course, if a tape recorder is not available, an additional pad of paper may be carried on the clipboard (under the checklist form) and these same notes can be written down rather than dictated.

As the inspector makes his rounds, photos are taken of all sides of the house. Also any other pertinent information is recorded on film. Such items might be ones that will be difficult to explain, verbally or in writing; thus, the photo will better help such persons to understand what you are talking about.

Figure 14-5: While this house may have basic Cape Cod features, other embellishments have been added. Will you have enough room on a checklist inspection form to note them?

After all items on the checklist have been covered, and adequate photos have been taken, a sketch of the house comes next. If original blueprints are available, the inspector merely checks (verifies) a few dimensions, and if they closely match the ones on the drawings, and no obvious additions have been made to the house, the inspector uses the set of original drawings and doesn't bother to take further measurements. However, if no drawings are available, the inspector should measure the outside dimensions of the house first; a sketch is then made of the house's outline (outside building lines). With this outline, he then moves to the interior of the home and sketches in the partitions and notes each room designation; that is, kitchen, living room, bath, bedroom, etc.

With the above information in hand, the inspector returns to his office for further work. The person we are discussing is rather "picky" and likes to do everything thoroughly. So with his list of items checked off at the job site, his tape recorder, and developed film, he then completes a blank-type inspection report form — like the one previously discussed. Both of these forms are read over several times, and while the project is fresh in his mind, he sits down at the computer and starts a rough draft on his final report to his clients. This latter (and final) report will be in the narrative format as discussed previously. Once the rough draft is completed, our inspector then proofreads this once or twice and corrects any errors found. He may also change around a few sentences while doing this. He then quits for the moment. The report is not yet ready to be turned over to his clients, but he stops to give his mind a break. The narrative report will be "polished" later. Besides, if he started out at, say, 8 o'clock in the morning, it should be time to break for lunch, and he's got another project to handle at 1 o'clock in the afternoon.

The Second Project

The afternoon project was actually on the same route as the first morning project. However, this particular inspector likes toget to a certain point on one project before going on to another. This way, the current project is fresh in his mind all the while — from the start of the inspection through a rough draft of the final report. Other inspectors may think this is a waste of time, and not too efficient, but we told you that this inspector is thorough — very thorough.

After arriving at the afternoon project, a similar inspection procedure is conducted as described for the morning project. When he looks at his watch after finishing up the rough draft for the afternoon project, it's almost 5 p.m. He then takes a break and finds himself in his garage, putting together a collection of fishing tackle for his planned weekend trip. About the time his tackle is collected, his wife calls him for the evening meal, after which he turns on the tv to catch the evening news. It's been a long day! He is tired from putting in nearly 9 hours of work, but he feels good about it. He completed two home inspections at $225 each, and did a thorough job on both. He knows his clients are going to be pleased. The thought of this puts him in a relaxed mood; so relaxed that he dozes in his easy chair for perhaps a total of 45 minutes. He awakens, refreshed, and the first thought that comes to mind is the final draft on his two reports. So back to the computer he goes.

Polishing The Report

The file for the morning project is brought up on the computer's monitor. He runs his spell-check program first. This picks up several errors made during the typing of his rough draft. He then runs the report through a grammer-check program that turns up a few more errors and instructs him on better sentence arrangement. The results are then carefully read — from start to finish — making a change here, another there, and then the ritual is repeated. After the third time around, this home inspector is satisfied that his report is the best work that he can do. It is then printed out in final form, the appropriate photos and a sketch of the house attached (along with his bill for the work), and all are put into an envelope for delivery to his client on the next day.

Before moving on to the afternoon project, he makes a back-up diskette for the report file, and files this diskette, along with his checklist and blank reports, left over photos, and a copy of his final report and invoice in the appropriate file in his filing cabinet. Reasons for this will be explained in the next section.

Our inspector then moves on to finalizing the report for the afternoon project. He goes through the exact same procedure as he did for the morning project, and then he's ready for his favorite tv shows before bed.

He has put in 12 hours work for the day, but by the same token, he has made nearly $500 during this time. He also has two more jobs the next day, but only one in the morning on Friday, so he'll have from noon Friday until Monday morning away from his business to engage in his hobby — fishing for lunker smallmouth bass in a nearby river.

Legalities

You can be the most qualified home inspector in the United States, do your work thoroughly, carefully, and on time. Still, you stand the chance of having a few customers or clients that are not satisfied. In some cases, your clients may refuse to pay you for your work or you could become involved in a lawsuit. This is where a good filing system and well prepared reports come in handy.

One way to prevent many dissatisfied customers and possible lawsuits is to have a pre-inspection contract with your client. Such a contract (Figure 14-6) tells the client what to expect and what the limitations will be. Still, some clients will threaten to bring suit as a means of renegotiating the cost of the inspection or the selling price of the house or to gain some other advantage. You will find such clients in any profession, so don't expect these individuals to be deterred by an inspector's professional excellence.

If you have done your job, and have evidence to prove it in the form of inspection reports (that's the reason you should make copies and backup computer diskettes of all your work), and photos, the home inspector can usually hold his ground and not be harmed by the courts.

If such a lawsuit arises, most attorneys believe that the narrative form of inspection report is the easiest for them to defend. Checklists and blank reports are usually too brief in critical areas and may force the parties to rely on the memory of who said what. Narratives are more likely to include all the results of the inspection and to have explanations for each finding.

If your inspection contract merges your oral statements into the written report, you should be careful not to contradict in writing what you said in person. Omissions can be critical, so be thorough. For these, and other reasons, the narrative report can be recommended over the others for your final report to your client. The checklist and blank report forms are fine for your on-site inspection, and for your files; but if possible, try to stick with the narrative style most of the time.

PRE-INSPECTION AGREEMENT
(PLEASE READ CAREFULLY)

COMPANY agrees to conduct an inspection for the purpose of informing the CUSTOMER of major deficiencies in the condition of the property. The inspection and report are performed and prepared for the sole, confidentials, and exclusive use and possession of the CUSTOMER. The written report will include the following only:

- structural condition and basement
- electrical, plumbing, hot-water heater, heating, and air conditioning
- quality, condition, and life expectancy of major systems

- general interior, including ceilings, walls, floors, windows, insulation, and ventilation
- kitchen and appliances
- general exterior, including roof, gutter, chimney, drainage, and grading

It is understood and agreed that this inspection will be of readily accessible areas of the building and is limited to visual observations of apparent conditions existing at the time of the inspection only. Latent and concealed defects and deficiencies are excluded from the inspection; equipment, items and systems will not be dismantled.

Maintenance and other items may be discussed, but they are not a part of our inspection. The report is not a compliance inspection or certification for past or present governmental codes or regulations of any kind.

The inspection and report do not address and are not intended to address the possible presence of or danger from any potentially harmful substances and environmental hazards including but not limited to radon gas, lead paint, asbestos, urea formaldehyde, toxic or flammable chemicals, and water, and airborne hazards. Also excluded are inspections of and reports on swimming pools, wells, septic systems, security systems, central vacuum systems, water softeners, sprinkler systems, fire and safety equipment, and the presence or absence of rodents, termites, and other insects.

The parties agree that the COMPANY, and its employees and agents, assume no liability or responsibility for the cost of repairing or replacing any unreported defects or deficiencies, either current or arising in the future, or for any property damage, consequential damage or bodily injury of any nature. THE INSPECTION AND REPORT ARE NOT INTENDED OR TO BE USED AS A GUARANTEE OR WARRANTY, EXPRESSED OR IMPLIED, REGARDING THE ADEQUACY, PERFORMANCE, OR CONDITION OR ANY INSPECTED STRUCTURE, ITEMS OR SYSTEM. COMPANY IS NOT AN INSURER OF ANY INSPECTED CONDITIONS.

It is understood and agreed that should COMPANY and/or its agents or employees be found liable for any loss or damages resulting from a failure to perform any of its obligations, including but not limited to negligence, breach of contract, or otherwise, then the liability of COMPANY and/or its agents or employees, shall be limited to a sum equal to the amount of the fee paid by the CUSTOMER for the inspection and Report.

Acceptance and understanding of this agreement are hereby acknowledged:

_____   _____
Company Representative   Date   Customer   Date

Figure 14-6: A pre-inspection agreement.

Index

A

Accounting
  billing and collecting 293
  cash flow 293
  payables, ...293
  statements 293
  system 290
Advertising 296
Agreement
  pre-inspection 316
Air conditioning
  concepts 235
  cycles 239
  humidity 238
  uses of 235
Ants
  carpenter 175
  identification of 173, 174
Attics
  definition of 104
  inspection of 109
  use of 30
  ventilation requirements . .104
Attire 14

B

Basements
  ceilings 147
  inspection of 67
Beams
  exposed 135
  grade 64
Bees
  identification 176
Blank reports
  definition of 303
  example of 304
Brick
  terms 79
  veneer 78, 80
Building setback 43
Building sites 39, 46, 48
Building structures
  types of 72, 73
  inspection of 81

C

Carpenter's level 81
Ceiling
  cathedral 137

coffered 141
definition of 132
drywall 145
finishes 133, 142
inspection of 84
insulation 148
plastered 133, 143
sagging 84
suspended 138, 146
tile 136
trayed 141
wood 135, 136
Checklists
bathroom 37
building sites and
  landscaping 55
ceilings 149
chimneys and flues 125
electrical 226
exteriors 26, 169, 170
foundation and basement ..22
house style 18
HVAC systems 252
insects, vermin and decay .185
interior 29
job 8
plumbing 274
roofs 110
structural 24, 86
Checklist reports
definition of 306
example of 307
Chimneys
cap 120
footing 120
hood 120
inspection of 114, 117
joints 119
leaks 119
test for leaking 115
use of 113

Circuits
electric 219
Ohm's law 220
branch 222
ground-fault interrupters ..225
Common power supplies ...193
Concrete block
layment 78
Concrete slabs 64, 65, 73
Construction
A-frame 76
balloon framing 76, 77
masonry 76
platform framing 76
pole 76
Contour map 42
Contours 53
Controls
HVAC 243
Crawl spaces
inspection of 66

D

Dampers 123, 124
Doors and windows
framing for 75
inspection of 83, 166
types of 27
Drain-waste-vent systems
fittings for 261
identification of 257
need for 260
terms 257, 258
Ductwork 248

E

Easements 42
Electrical system
inspection 34, 222

Index

secondary 192
underground 191
Electricity
  generation 187
  services 222
  transmission 190
Electric space heating
  heaters 250
  types of 233
  use of 232
Equipment
  for inspectors 10-14
  heating/cooling 233
Erosion
  soil 53
Exterior treatment
  inspection 23
  wood finishes . .151, 152, 153

F

Field inspection 15
Fireplace
  damper 122, 123
  inspection of 122
  lintel 122, 123
Flashing
  definition of 99
  where installed 100, 101
Fleas
  identification of 179
  treatment for 178
Floors
  inspection of 81
Flue
  inspection of 118
Footings
  building codes 58
  dimensions 58
  material made of 57
Foundations 19, 57

Foundation walls
  anchors 62
  inspection of 66
  materials 60, 61
Frostline 59

G

Ground-fault circuit
  interrupter 225
Gypsum board
  compound 144
  definition of 134
  installation of 134

H

Home inspection
  business 275
  getting started 286
  income 282, 283
  need for 276
  office setup 286, 289
  partner 284
  self-checklist 277
  starting cost 280
  tools 281
Homes
  evaluation of 45
  style 15
Hot water heaters
  components 262
  elements 264
  indirect heaters 268
  installation 265
  operation 263, 265
  protection 268
  thermostats 265
Humidity 238
HVAC systems
  controls 243

definition of 227
inspection 32, 246

I

Insects
  ants 174, 175
  bees and wasps 176
  fleas 178
  flying ant 173
  inspection for 182
  roaches 176
  termites 172, 173
Interior treatment
  identification of finish 129
  inspection of 25, 128

K

Kilowatt hours 189
Kilowatts 189

L

Landscaping
  definition of 48
  drainage 49, 50, 51, 52
  inspection of 54
Legalities of business 315

N

Narrative reports
  definition of 299
  example of 300, 301, 302
National Electrical Code
  address of 198
  articles 207, 208, 209, 210
  definitions 214
  identification of 197
  layout 202, 203, 204, 205
  practical
    applications ...212, 213, 214
  terminology 199
  use of 210
National Electrical
  Manufacturers Association .218
National Fire Protection
  Association 218
Nationally Recognized
  Testing Laboratory 218
Neighborhood development .47

O

Office
  activities 290
  layout 289
  setup 286
Ohm's law 219
Outside coverings
  inspection of 168

P

Piers
  definition of 63
  types 63
Pipes
  air chambers 256
  hot water 266
  insulation of 256
Plaster
  how to mix 133
  principles of 134
Plot plan 44
Plumbing
  basic system 253, 254
  inspection 270, 271
  purpose of 253
  traps 258
  vents 259

Q

Questions for homeowner . . .28

R

Radon gas
  definition of 181
  entrance of 183
  prevention 181, 182
  testing for 181
Rafters
  common 102
  cripple 103
  hip 102
  jack 103
  valley 102
Records
  bookkeeping 290
  choice of 292
  need for 291
Relations
  community 295
  customer 294
  cublic 294
Relief valves 267
Reports
  finishing of 314
  practical application 308
  preparation of 297
  types of 298
Roaches
  habitat 176
  identification 177
  treatment 177
Roofing
  asphalt roll 97
  built-up 97, 107
  corrugated 98
  fiberglass 99
  inspection of 108
  metal 98
  slate 97
  structural metal batten
    system 99
Roofs
  definition of 87
  flashing 99, 100
  inspection of 104, 105
  material for 93-98
  metal 108
  pitches 92
  terms for 90, 91
  tile 108
  types of 88, 89, 90

S

Shingles
  aluminum shakes 95
  asphalt 94, 106
  clay-tile 95
  shake applications 156, 17
  wood 93, 107, 158
Siding
  aluminum 159, 165
  asbestos-cement 158
  bevel 152
  board and batten 155
  drop 153
  horizontal 153, 154
  log 160
  masonry 164
  plywood 155
  stucco 161
  vertical 153, 154
  vinyl 159, 165
Soil
  condition of 5
  drainage 49
  sample 46

Structural members 21
Structures
  masonry 72
  prefabricated 72, 80
  reinforced concrete 72
  steel frame 72
  structural steel 72
  wood-frame 72
Swales 47
Systems
  air conditioning 235
  DWV 256, 257, 260
  forced warm air 228
  hot water 262
  hot water heating 229
  HVAC 228
  plumbing 255
  sewer-septic 261
  water supply 256

T

Termites
  identification 173
  shield 173
  tunnels 172
Traps
  cleanouts for 259
  definition 258
  in sinks 258, 259
  in water closet 258
  need for 260
  vents for 259
Treads on stairs
  inspection of 129, 130
Trim
  definition 130
  inspection of 130
  outside 167
Trusses
  definition of 103

U

Underwriters' Laboratory
  definition of 217
Utilities 53

V

Veneer
  brick 78, 80
Ventilation
  problems of 35
Vermin
  definition of 179
  inspection for 182
  prevention 179
Vibration
  testing for 83

W

Wainscoting
  use of 131
Walls
  plaster 127
  papered 127
Water table 46
Wood construction
  consists of 73
  decay 172, 180
  floor framing details 74
  roofs 74, 75
  types of 74
  walls 73
Wood stoves
  inspection 116
  installation 118
  stovepipe for 116
Wythe 121

Z

Zoning laws 41
Zoning variance 42

Practical References for Builders

Renovating & Restyling Older Homes

Any builder can turn a run-down old house into a showcase of perfection — if the customer has unlimited funds to spend. Unfortunately, most customers are on a tight budget. They usually want more improvements than they can afford — and they expect you to deliver. This book shows how to add economical improvements that can increase the property value by two, five or even ten times the cost of the remodel. Sound impossible? Here you'll find the secrets of a builder who has been putting these techniques to work on Victorian and Craftsman-style houses for twenty years. You'll see what to repair, what to replace and what to leave, so you can remodel or restyle older homes for the least amount of money and the greatest increase in value.
416 pages, 8^1/$_2$ x 11, $33.50

National Repair & Remodeling Estimator

The complete pricing guide for dwelling reconstruction costs. Reliable, specific data you can apply on every repair and remodeling job. Up-to-date material costs and labor figures based on thousands of jobs across the country. Provides recommended crew sizes; average production rates; exact material, equipment, and labor costs; a total unit cost and a total price including overhead and profit. Separate listings for high- and low-volume builders, so prices shown are specific for any size business. Estimating tips specific to repair and remodeling work to make your bids complete, realistic, and profitable. Includes a CD-ROM with an electronic version of the book with *National Estimator*, a stand-alone *Windows*™ estimating program, plus an interactive multimedia video that shows how to use the disk to compile construction cost estimates.
296 pages, 8^1/$_2$ x 11, $48.50. Revised annually

National Renovation & Insurance Repair Estimator

Current prices in dollars and cents for hard-to-find items needed on most insurance, repair, remodeling, and renovation jobs. All price items include labor, material, and equipment breakouts, plus special charts that tell you exactly how these costs are calculated. Includes a CD-ROM with an electronic version of the book with *National Estimator*, a stand-alone *Windows*™ estimating program, plus an interactive multimedia video that shows how to use the disk to compile construction cost estimates.
568 pages, 8^1/$_2$ x 11, $49.50. Revised annually

CD Estimator

If your computer has *Windows*™ and a CD-ROM drive, *CD Estimator* puts at your fingertips 85,000 construction costs for new construction, remodeling, renovation & insurance repair, electrical, plumbing, HVAC and painting. You'll also have the *National Estimator* program — a stand-alone estimating program for *Windows*™ that *Remodeling* magazine called a "computer wiz." Quarterly cost updates are available at no charge on the Internet. To help you create professional-looking estimates, the disk includes over 40 construction estimating and bidding forms in a format that's perfect for nearly any word processing or spreadsheet program for *Windows*™. And to top it off, a 70-minute interactive video teaches you how to use this CD-ROM to estimate construction costs. **CD Estimator is $68.50**

Illustrated Guide to the 1999 *National Electrical Code*

This fully-illustrated guide offers a quick and easy visual reference for installing electrical systems. Whether you're installing a new system or repairing an old one, you'll appreciate the simple explanations written by a code expert, and the detailed, intricately-drawn and labeled diagrams. A real time-saver when it comes to deciphering the current *NEC*. **360 pages, 8^1/$_2$ x 11, $38.75**

Home Inspection Illustrated Detailing Manual

This is probably the most complete how-to guide available today on finding, identifying, and fixing problem areas in homes. It's filled with hundreds of illustrations that show what can go wrong with a house and how the problem can usually be fixed. Every area of the house is covered: structural, exterior, roofs, plumbing, electrical, heating, air conditioning, interior, kitchen, baths and utility areas.
304 pages, 8¹/2 x 11, spiral bound, $34.95.
Published by Builder's Book, Inc.

Contractor's Guide to QuickBooks Pro 2000

This user-friendly manual walks you through QuickBooks Pro's detailed setup procedure and explains step-by-step how to create a first-rate accounting system. You'll learn in days, rather than weeks, how to use QuickBooks Pro to get your contracting business organized, with simple, fast accounting procedures. On the CD included with the book you'll find a QuickBooks Pro file preconfigured for a construction company (you drag it over onto your computer and plug in your own company's data). You'll also get a complete estimating program, including a database, and a job costing program that lets you export your estimates to QuickBooks Pro. It even includes many useful construction forms to use in your business. **304 pages, 8^1/$_2$ x 11, $44.50**

Profits in Buying & Renovating Homes

Step-by-step instructions for selecting, repairing, improving, and selling highly profitable "fixer-uppers." Shows which price ranges offer the highest profit-to-investment ratios, which neighborhoods offer the best return, practical directions for repairs, and tips on dealing with buyers, sellers, and real estate agents. Shows you how to determine your profit before you buy, what "bargains" to avoid, and how to make simple, profitable, inexpensive upgrades. **304 pages, 8^1/$_2$ x 11, $19.75**

Craftsman Book Company
6058 Corte del Cedro, P.O. Box 6500
Carlsbad, CA 92018

☎ 24 hour order line
1-800-829-8123
Fax (760) 438-0398

Order online
http://www.craftsman-book.com
Free on the Internet! Download any of Craftsman's estimating costbooks for a 30-day free trial! http://costbook.com

Name _____
e-mail address (for special offers) _____
Company _____
Address _____
City/State/Zip _____
○ This is a residence
Total enclosed_____ (In California add 7.25% tax)

We pay shipping when your check covers your order in full.

In A Hurry?
Use your ○ Visa ○ MasterCard
○ Discover or ○ American Express
Card _____
Exp. date _____ Initials _____

Tax Deductible: Treasury regulations make these references tax deductible when used in your work. Save the canceled check or charge card statement as your receipt.

10-Day Money Back Guarantee

○ 68.50 CD Estimator
○ 44.50 Contractor's Guide to QuickBooks Pro 2000
○ 34.95 Home Inspection Illustrated Detailing Manual
○ 38.75 Illustrated Guide to the 1999 *National Electrical Code*
○ 49.50 National Renovation & Insurance Repair Estimator with FREE *National Estimator* on a CD-ROM.
○ 48.50 National Repair & Remodeling Estimator with FREE *National Estimator* on a CD-ROM.
○ 19.75 Profits in Buying & Renovating Homes
○ 33.50 Renovating & Restyling Older Homes
○ 24.95 Home Inspection Handbook
○ FREE Full Color Catalog

Prices subject to change without notice

Craftsman Book Company
6058 Corte del Cedro, P.O. Box 6500
Carlsbad, CA 92018

☎ 24 hour order line
1-800-829-8123
Fax (760) 438-0398

_____
Name
_____
e-mail address (for special offers)
_____
Company
_____
Address
_____
City/State/Zip
○ This is a residence

Total enclosed_____(In California add 7.25% tax)

We pay shipping when your check covers your order in full.

In A Hurry?
Use your ○ Visa ○ MasterCard
○ Discover or ○ American Express

Card#_____

Exp. date_____Initials_____

Tax Deductible: Treasury regulations make these references tax deductible when used in your work. Save the canceled check or charge card statement as your receipt.

Order online
http://www.craftsman-book.com
Free on the Internet! Download any of Craftsman's estimating costbooks for a 30-day free trial! http://costbook.com

10-Day Money Back Guarantee

○ 68.50 CD Estimator
○ 44.50 Contractor's Guide to QuickBooks Pro 2000
○ 34.95 Home Inspection Illustrated Detailing Manual
○ 38.75 Illustrated Guide to the 1999 *National Electrical Code*
○ 49.50 National Renovation & Insurance Repair Estimator with FREE *National Estimator* on a CD-ROM.
○ 48.50 National Repair & Remodeling Estimator with FREE *National Estimator* on a CD-ROM.
○ 19.75 Profits in Buying & Renovating Homes
○ 33.50 Renovating & Restyling Older Homes
○ 24.95 Home Inspection Handbook
○ FREE Full Color Catalog

Prices subject to change without notice

Mail This Card Today
For a Free Full Color Catalog

Over 100 books, annual cost guides and estimating software packages at your fingertips with information that can save you time and money. Here you'll find information on carpentry, contracting, estimating, remodeling, electrical work, and plumbing.

All items come with an unconditional 10-day money-back guarantee. If they don't save you money, mail them back for a full refund.

_____
Name
_____
e-mail address (for special offers)
_____
Company
_____
Address
_____
City/State/Zip

Craftsman Book Company / 6058 Corte del Cedro / P.O. Box 6500 / Carlsbad, CA 92018

NO POSTAGE
NECESSARY
IF MAILED
IN THE
UNITED STATES

BUSINESS REPLY MAIL
FIRST CLASS MAIL PERMIT NO. 271 CARLSBAD, CA

POSTAGE WILL BE PAID BY ADDRESSEE

 Craftsman Book Company
6058 Corte del Cedro
P.O. Box 6500
Carlsbad, CA 92018-9892

Il.l....l.lll......lll..l.l.l..l..l.l.l....l.l..l.ll

NO POSTAGE
NECESSARY
IF MAILED
IN THE
UNITED STATES

BUSINESS REPLY MAIL
FIRST CLASS MAIL PERMIT NO. 271 CARLSBAD, CA

POSTAGE WILL BE PAID BY ADDRESSEE

 Craftsman Book Company
6058 Corte del Cedro
P.O. Box 6500
Carlsbad, CA 92018-9892

Il.l....l.lll......lll..l.l.l..l..l.l.l....l.l..l.ll